PROCEEDINGS OF THE SECOND INTERNATIONAL CONFERENCE ON TAILINGS
& MINE WASTE '95/FORT COLLINS/COLORADO/USA/17-20 JANUARY 1995

Tailings &
Mine Waste '95

A.A.BALKEMA/ROTTERDAM/BROOKFIELD/1995

Organized by Geotechnical Engineering Program, Department of Civil Engineering, Colorado State University

CONFERENCE ORGANIZING COMMITTEE

John D. Nelson, Chair   Jim Pendleton
Robert L. Medlock   Harry H. Posey
Debora J. Miller   Charles D. Shackelford
Louis L. Miller   Dirk J. A. Van Zyl
Janet Lee Montera

The texts of the various papers in this volume were set individually by typists under the supervision of each of the authors concerned.

Published by
A.A. Balkema, P.O. Box 1675, 3000 BR Rotterdam, Netherlands (Fax: +31.10.4135947)
A.A. Balkema Publishers, Old Post Road, Brookfield, VT 05036, USA (Fax: 802.2763837)

ISBN 90 5410 526 7
© 1995 A.A. Balkema, Rotterdam
Printed in the Netherlands

Table of contents

Preface   VII

Collection lysimeters for in situ monitoring of compacted clay liners   1
Te-Fu Chiu

Abandoned uranium mine reclamation – A case study   11
E.T.Cotter

Gypsum may delay acid production from waste rock   19
L. H. Filipek, T.J.VanWyngarden, T.R.Wildeman, C.S.E. Papp & J.Curry

Geochemical modeling of waste rock leachate generated in precious metal   23
mines
J. P. Kaszuba, W.J. Harrison & R. F.Wendlandt

Response to the EPA four focused feasibility studies regarding the Summitville 35
mine clean-up by the Technical Assistance Group
D. Kent, J. Stern, T.Gilmer, K. Klco, M. Mueller, M.Ter Kuile, C. Canaly &
W.Mellott

Metal transport between an alluvial aquifer and a natural wetland impacted   43
by acid mine drainage, Tennessee Park, Leadville, Colorado
S. S. Paschke & W.J. Harrison

Superfund listing of mining sites   55
C.A. Patton, K. M. McGaffey, J. L. Ehrenzeller, R. E. Moran & W.S. Eaton

Simulation of pit closure alternatives for an open pit mine   67
H. F. Pavlik, F.G. Baker, Xiaoniu Guo & J. S.Voorhees

Inventorying and characterization of mine wastes through remote sensing: 79
The Cripple Creek mining district
D. C. Peters, P. L. Hauff & K. E. Livo

Recovery of water quality in a mine pit lake after removal of aqua-cultural   81
waste loading: Model predictions versus observed changes
H. M. Runke & C. J. Hathaway

Analysis of tailings and mine waste sediment transport in rivers   91
R. K. Simons, D. B. Simons & G. E. Canali

Stream restoration and mine waste management along the upper Clark Fork   105
River
C. T. Stilwell

Prediction of tailings effluent flows   109
D. E. Welch, J. M. Johnson & L. C. Botham

Integrated design of tailings basin seepage control systems using analytic   121
element ground-water models
R. W. Wuolo & P. E. Nemanic

Author index   131

Tailings & Mine Waste'95 © 1995 Balkema, Rotterdam, ISBN 90 5410 526 7

Preface

In 1978, the first Symposium on Uranium Mill Tailings Management was held at Colorado State University. Over the following 10 years or so, this symposium became popular in that it provided a forum for regulators, operators, researchers, and consultants involved in uranium mill tailings management to discuss issues of common interest. The symposium embraced a broad range of topics and a broad spectrum of technical expertise. It provided the opportunity for the entire uranium mining industry to keep abreast of the state of the art, new technology and ideas, regulatory changes, and issues of overall general interest.

When the uranium market declined and uranium mills closed, the support for this symposium waned. However, there has remained a need for a forum of this nature to address issues related to tailings management for the entire mining industry as a whole. Consequently, Colorado State University initiated a conference on tailings and mine waste in 1994. That conference was well-attended and met with success. This conference, Tailings & Mine Waste'95, is the second of these. It is also being held in conjunction with the Summitville Forum. We are confident that these two events will be of great interest and value to the mining community.

Organizing Committee

Tailings & Mine Waste'95 © *1995 Balkema, Rotterdam, ISBN 90 5410 526 7*

Collection lysimeters for in situ monitoring of compacted clay liners

Te-Fu Chiu
Department of Civil Engineering, Colorado State University, Ft. Collins, Colo., USA

ABSTRACT: A summary of the geotechnical and hydrological features of collection lysimeters as well as the measurement of the in situ hydraulic conductivity of compacted clay liners using collection lysimeters is presented. The significance of the capillary barrier effect in terms of collection lysimeters for compacted clay liners is illustrated. Five existing landfills and two prototype cells where collection lysimeters are used as the monitoring system are reviewed. The limitations in the use of collection lysimeters for monitoring purposes are addressed.

INTRODUCTION

Over the last ten years, compacted clay liners (CCLs) have received considerable attention as hydraulic barriers in water retention ponds, waste disposal pits, landfills, and other impoundments. Most criteria for the design and construction of CCLs are based on achieving a low value of the saturated hydraulic conductivity, K_s (e.g., $K_s \leq 1$ x 10^{-7} cm/s). This basis is sound and can be achieved in a laboratory when a appropriate soil is compacted by a suitable compaction method. However, Daniel (1984) reports that the field hydraulic conductivity of compacted clay liners can be 10 to 1,000 times greater than the laboratory measured hydraulic conductivity. Additional studies conducted by Day and Daniel (1985) concluded that laboratory tests underestimate the actual hydraulic conductivity and greater emphasis should be placed on field tests. Although four types of in situ hydraulic conductivity tests are recommended, only collection lysimeters and infiltrometers possess the ability to permeate a relative large volume of soil (Daniel, 1989).

Collection lysimeters, also known as collection pans, pan lysimeters, and underdrains, are lined basins or troughs constructed beneath or within CCLs. The lysimeter typically is backfilled with drainage material, soaked, and covered with a fabric filter. In general, the collection lysimeters are required to be tested for leaks for 24 to 48 hours. A typical cross section of a collection lysimeter is shown in Fig. 1 in which a perforated pipe extends from the bottom of the basin to an access point. The collected liquid can be extracted and the quantity as well as quality of the liquid can be determined. As a monitoring tool, collection lysimeters provide not only data to determine the in situ hydraulic conductivity of CCLs but also an early warning for the long-term performance of a landfill.

Collection lysimeters as described above were first used at the Nekoosa Papers sludge landfill, Wisconsin, in 1976 (Kmet and Lindorff, 1983). Data from facilities with collection lysimeters are available in the literature (Kmet and Lindorff, 1983; Gordon et al., 1989; Reades et al., 1987, 1990). However, the data reported thus far pertain to a relatively short-term period (e.g., eight years for the longest) relative to the design life of

1

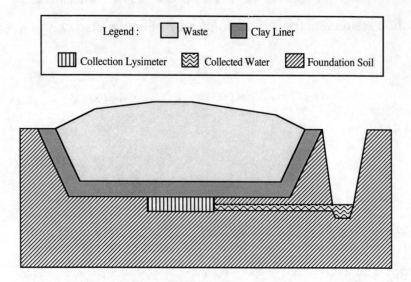

Figure 1 - Schematic Cross-Section of Collection Lysimeter for Measurement of Field
Hydraulic Conductivity (after Chiu and Shackelford, 1994)

a landfill, and most of the in situ hydraulic conductivity measurements based on the data
have been lower than the laboratory measured values. However, these field-measured
hydraulic conductivity values may be relatively low initially due to unsaturated flow
behavior relative to saturated flow behavior, in general, and the capillary barrier effect, in
particular (Shackelford et al, 1994; Chiu and Shackelford, 1994).

   The purpose of this paper is to illustrate the influence of the capillary barrier effect on
the measurement of in situ hydraulic conductivity of CCLs when collection lysimeters are
used as the monitoring system. After reviewing the observation data of five existing
landfills and two prototype cells, the limitation in the use of collection lysimeters are
addressed.

CAPILLARY BARRIER EFFECT

One-dimensional unsaturated flow through a fine-grained soil overlying a coarse-grained
soil may result in what is commonly referred to as "the capillary barrier effect." Due to the
different soil-moisture characteristics of the two soils, only a fraction of the water flux
reaching the interface of the two soils is transmitted into the coarse-grained soil. The
remaining portion of the infiltrating water flux is "reflected" upward into the overlying
fine-grained soil due to the residual suction remaining in the fine-grained soil.

   In order to depict the capillary barrier effect, the wetting soil-moisture characteristic
curves of two soils are schematically illustrated in Fig. 2 in terms of capillary pressure
head, h_c, and normalized volumetric water content, θ^*. Point (a) in Fig. 2 represents the
capillary pressure head on the fine-grained soil side just when the water flux reaches the
interface between the two soil layers. Both the capillary pressure head and the water flux,
based on Darcy's law, must be the same for the two soils at the interface due to the
requirement for continuity across the interface. As a result, the capillary pressure head on
the coarse-grained soil side must be at point (a') to satisfy the continuity of capillary

2

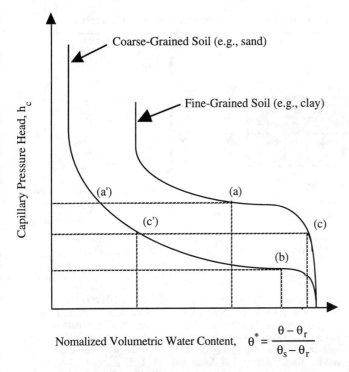

Legend:
θ   = volumetric water content
θ$_r$ = residual volumetric water content
θ$_s$ = saturated volumetric water content

Coarse-Grained Soil (e.g., sand)

Fine-Grained Soil (e.g., clay)

Capillary Pressure Head, h$_c$

(a')   (a)

(c')   (c)

(b)

Nomalized Volumetric Water Content, $\theta^* = \dfrac{\theta - \theta_r}{\theta_s - \theta_r}$

Figure 2 - Schematic Wetting Soil-Moisture Characteristic Curves for Two Soils (after Morel-Seytoux, 1992a,b; Shackelford et al., 1994; Chiu and Shackelford, 1994)

pressure. However, due to the continuity of water flux, the water content on the coarse-grained soil side must be at point (b) which is greater than the incipient value. Since the capillary pressure heads at points (a') and (b) do not represent the same values, the two requirements of continuity cannot be satisfied if the water content in the fine-grained soil side remains at point (a) while the water content in the coarse-grained soil side increases. As a result, the water content of both soils must increase simultaneously to points (c) and (c') so that the capillary pressure head will be the same. The physical significance is that the water flux is not completely transmitted into the coarse-grained soil, i.e., a fraction of the water flux must "reflect" upward into the fine-grained soil.

Practical examples of the capillary barrier effect include compacted clay covers underlain by sand drainage layers, compacted clay liners underlain by leachate collection/removal systems, and collection lysimeters used as monitoring systems beneath compacted clay liners (Chiu and Shackelford, 1994). Based on the numerical simulation results conducted by Chiu and Shackelford (1994), for the scenarios where the width, W, of coarse-grained

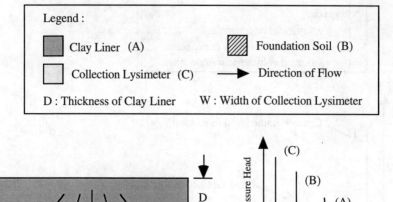

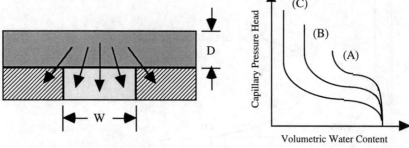

Figure 3 - Two-Dimensional Flow Due to the Difference of Soil-Moisture Characteristic Curves of Soils (modified after Chiu and Shackelford, 1994)

soil is large relative to the thickness, D, of the fine-grained soil (e.g., W/D≥10), a one-dimensional capillary barrier effect may be approximated in practice. However, if this criterion is not met, a two-dimensional unsaturated flow may result. This situation may occur when the width of a collection lysimeter is significantly less than the thickness of the compacted clay liner. In this case, the water flux will flow laterally around the collection lysimeter into the foundation soil as illustrated in Fig. 3.

Factors affecting the capillary barrier effect have been studied by Morel-Seytoux (1992a,b), Shackelford et al. (1994), and Chiu and Shackelford (1994). These factors, including the initial conditions, soil-moisture characteristics, and the saturated and unsaturated hydraulic conductivity of the soils combined with the boundary conditions, infiltration rate, and size of the barriers, will result in different degrees of capillary barrier effect. Typical breakthrough curves shown in Fig. 4 represent the relationship between flux ratio, which is the ratio of transmitted flux versus infiltration flux, and elapsed time. The influence of initial saturation of the compacted clay and the size ratio, W/D, are illustrated in Fig. 4. A time-lag is usually expected before the full water flux can be transmitted into the underlying soil due to the capillary barrier effect.

CASE STUDIES

In order to provide practical examples of the potential effect of unsaturated flow on the in situ measurement of the hydraulic conductivity of CCLs, five existing landfills and two prototype sites where collection lysimeters are used as monitoring systems are reviewed. The observation data as well as some hydrogeological features of those landfills are summarized in Table 1.

4

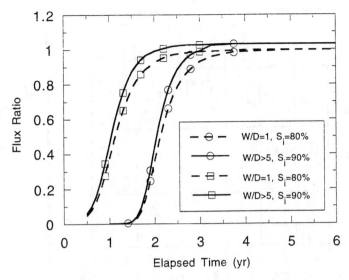

Figure 4 - Typical Breakthrough Curves of Water Flux Showing the Influences of Initial Saturation, S_i, of Compacted Clay and Size Ratio, W/D (after Chiu and Shackelford, 1994)

One-dimensional saturated flow has been assumed in all reported cases and a saturated form of Darcy's law has been used in the calculation of in situ hydraulic conductivity. At four of the five landfill sites, the in situ hydraulic conductivities of the CCLs decrease from the installation dates to values lower than the laboratory results (no lab data are available for Nekoosa Papers landfill). For the extreme cases, no flow was detected after one year of operation (12/80 to 12/81) for the six lysimeters of Marathon County landfill in Wisconsin, and these lysimeters were still dry as of the latest reported date (12/87). Similarly, at Keele Valley landfill, Canada, two of the six lysimeters buried within the CCLs observe no flow after 1989 (five years after installation). In addition, all five lysimeters at the Keele Valley landfill located beneath the CCLs have remained dry since the construction date. In some cases, the measured hydraulic conductivity values are larger at relatively early times. These relatively large hydraulic conductivity values can be attributed to collection of large amounts of water in the early stage due to consolidation of the CCLs (Bonaparte and Gross, 1990). The decrease in hydraulic conductivity with time has been explained on the basis of a reduction of the in situ hydraulic conductivity due to the increase of overburden and the clogging of soil pores (Reades et al., 1990). However, the W/D ratios and the contrast in soil-moisture characteristics between backfill material and foundation soils also may explain these observations in terms of the capillary barrier effect. Based on Chiu and Shackelford (1994), a smaller W/D ratio will result in a greater capillary barrier effect which is the situation found at the Marathon County landfill (W/D ranges from 3 to 4.5). On the other hand, a high contrast in soil-moisture characteristics, including saturated and unsaturated hydraulic conductivity, between backfill material and foundation soils also will result in a greater capillary barrier effect. Although there is not enough information provided to quantify these potential effects, the contrast between a medium to coarse sand and a dense silty sand with 20% fines on Marathon County site could be relatively high. Similarly, at the Keele Valley site, the contrast in soil-moisture characteristics between sand lysimeters and a 1 - to 3 - m thick sandy layer with a water table located at the bottom of the layer also can be high. In addition, for the six sand

Table 1. Summary of the Performance of Some Landfills Using Collection Lysimeters

Landfill	D	S_i (%)	H_L	Size of Lysimeter	Backfill Material	Foundation Soils	Field K (cm/s)	Elapsed Time	Lab K (cm/s)	Field K / Lab K	Reference
Nekoosa Papers	None	NA	NA	10x10x2 ft^3	On-Site Sand	Fine to Coarse Sand	3 x 10^{-9} ('76) to 2 x 10^{-7} ('83)	4 - 7 yr	NA	NA	Kmet and Lindorff (1983)
	4 ft	NA	NA	1.25x20 ft^2			1 x 10^{-7} ('81) to 4 x 10^{-7} ('83)	3 yr	NA	NA	
Marathon County	4 ft	NA	NA	18x40x3 ft^3	Medium to Coarse Sand	Dense Silty Sand with 20% -No.200	2 x 10^{-8} (12/80) Dry after 12/81	8 yr	3 x 10^{-9}	6.7 0	Kmet and Lindorff (1983); Gordon et al. (1989)
	4 ft	NA	<3ft	12x100x3 ft^3			5 x 10^{-9} (9/85) Dry after 12/86	4 yr	3 x 10^{-9}	1.6 to 0	
Portage County	5 ft	NA	NA	7400 ft^2	NA	Sandy Glacial Till with 35% -No.200	9 x 10^{-9} (7/84) to 1 x 10^{-9} (12/87)	3 yr	1 x 10^{-8}	0.9 to 0.1	Gordon et al. (1989)
Sauk County	5 ft	NA	NA	900 ft^2	NA	Fine to Medium Sand	4 x 10^{-8} (9/83) to 1 x 10^{-8} (12/87)	4 yr	8 x 10^{-8}	0.5 to 0.125	Gordon et al. (1989)
Keele Valley, Maple, Ontario, Canada	0.3 - 0.5 m	95	1 m	15x15 m^2 within CCL	Sand	1 - 3 m thick sand	1 x 10^{-7} (3/84) to 1 x 10^{-9} (3/88)	4 yr	3 x 10^{-9} to 2 x 10^{-8}	5 to 33.3 and 0.1 to 0.33	Reades et al. (1987, 1990)
	1.2 m	95	1 m	15x15 m^2 below CCL			No Flow (5/90)	6 yr	3 x 10^{-9} to 2 x 10^{-8}	0	

Table 1. (Cont'd)

Landfill	D	S_i (%)	H_L	Size of Lysimeter	Backfill Material	Foundation Soils	Field K (cm/s)	Elapsed Time	Lab K (cm/s)	Field K / Lab K	Referenc
Prototype Cells	6 in	92	<6ft	NA	3/4 in Gravel	Lined with 40-mil HDPE	9×10^{-6}	12 weeks	1×10^{-8} to 1×10^{-7}	90 to 900	Day and Daniel (1985)
	6 in	90	<6ft	NA			4×10^{-6}	12 weeks	2×10^{-9} to 4×10^{-8}	100 to 2,000	
Protptype Cells	1 ft	NA	1 ft	256 ft^2	Gravel	NA	1×10^{-4}	22 days	6×10^{-10} to 1×10^{-7}	1,000 to 166,666	Elsbury et al. (1990)

where D = thickness of CCL
S_i = initial saturation of CCL
H_L = leachate head perched on the CCL

7

lysimeters buried within the CCLs at Keele Valley landfill, the two-dimensional capillary barrier effect as described by Chiu and Shackelford (1994) could be dominating the unsaturated flow.

One exceptional case is the Nekoosa Papers landfill, Area II, where the in situ hydraulic conductivity increases steadily during the three-year observation period. No periodic observation data are provided in the report; however, the increase of the measured flow is similar to the predicted breakthrough curve (refer to Fig. 4). Although the W/D ratio is very small (W/D = 5/16), the collection lysimeter is backfilled with on-site sand. As a result, the contrast of soil-moisture characteristic between lysimeter and foundation soil is reduced and the water flux would be similar to the one-dimensional rather than two-dimensional scenario.

Problems involving the capillary barrier effect can be complicated in practical situations. All of the factors listed in Table 1 may affect each other and contribute more or less to the in situ monitoring results. The thickness, D, and the initial saturation, S_i, of CCLs combined with the leachate head will control the infiltration rate. The size of the lysimeter, the soil-moisture characteristics of backfill material and foundation soil, together with the position of water table determine the boundaries of the problem. One factor not listed in Table 1 is the observation period. All of the observation periods in the reports are relatively short (≤ 8 years) compared with the design life of a typical landfill. Based on the water flux breakthrough curve shown in Fig. 4, a significant water flux in the collection lysimeters may not be apparent during the time-lag period. The length of the time-lag as well as the slope of the S-shape breakthrough curve are influenced by the factors mentioned above. All of these factors are not discussed in the reports referenced in Table 1. Actually, a one-dimensional flow and Darcy's saturated equation are simply assumed and used in the calculations. When the lysimeters detect no flow, clogging of pores is always the first conclusion. As a result, one disadvantage in using collection lysimeter is the difficulty of detecting when problems occur.

Finally, the performance of two prototype tests are considered (Day and Daniel, 1985; Elsbury et al., 1990). The two-dimensional influence of capillary barrier effect can be neglected since no foundation soil is in contact with the CCL in the tests by Day and Daniel (1985), and the size of lysimeter is almost the same as the testing pond for the tests conducted by Elsbury et al. (1990). In both of these cells, field hydraulic conductivity measured by collection lysimeter are close to the results of infiltrometer tests. However, the in situ values are much higher (1 to 5 orders of magnitude) than the laboratory results which is conflict with all other observations. Poor construction quality control may have resulted in the relatively large field values of hydraulic conductivity (Reades et al., 1989). In addition, the thickness of CCLs used in both prototype cells are much less than the thickness of typical soil liners. Thus, it is apparent that one possible explanation for the lack of an apparent capillary barrier effect in these tests is the existence of secondary features or macropores which short circuit the flow system.

CONCLUSIONS

The use of collection lysimeters within or beneath compacted clay liners (CCLs) results in a potential capillary barrier effect. Due to the capillary barrier effect, one-dimensional flow may not exist and the use of saturated Darcy's equation may underestimate the in situ hydraulic conductivity of CCLs. Many factors may contribute to this underestimation including the thickness and initial saturation of CCLs, the size of collection lysimeter, the soil-moisture characteristics of the backfill materials used for the collection lysimeter and foundation soils, the leachate head perched on the top of liner, and the elapsed time of observation.

The performance of five existing landfills and two prototype liners are reviewed in this paper. All of the case histories ignore the capillary barrier effect and represent relatively short observation periods. Since the capillary barrier effect results in a reduced water flux relative to the saturated assumption in the short term, the evaluation of the monitoring data for these case histories may be not only incorrect but also unconservative; the capillary barrier effect should be considered when interpreting data obtained from collection lysimeters for the purpose of measuring the in situ hydraulic conductivity of compacted clay liners.

REFERENCES

Bonaparte, R. and B.A. Gross 1990. Field Behavior of Double-Liner Systems. *Waste Containment System: Construction, Regulation, and Performance*. R. Bonaparte Ed., Geotechnical Special Publication, No. 26, Nov. 1990.
Chiu, Te-Fu and C.D. Shackelford 1994. Practical Aspect of the Capillary Barrier Effect for Landfill. *Proceedings of the Seventeenth International Madison Waste Conference, Sep. 21-22, 1994, Dept. of Engrg. Professional Development, University of Wisconsin* .
Daniel, D.E. 1984. Predicting Hydraulic Conductivity of Clay Liners. *J. of Geotechnical Engrg.*, ASCE, 110(2): 285-300.
Daniel, D.E. 1989. In Situ Hydraulic Conductivity Tests for Compacted Clay. *J. of Geotechnical Engrg.*, ASCE, 115(9): 1205- 1226.
Day, S.R. and D.E. Daniel 1985. Hydraulic Conductivity of Two Prototype Clay Liners. *J. of Geotechnical Engrg.*, ASCE, 111(8): 957-970.
Elsbury, B.R., G.A. Sraders, D.C. Anderson, J.A. Rehage, J.O. Sai, and D.E. Daniel 1990. Project Summary: Field and Laboratory Testing of a Compacted Soil Liner. *EPA/600/S2-88/067*.
Gordon, M.E., P.M. Huebner, and T.J. Miazga 1989. Hydraulic Conductivity of Three Landfill Clay Liners. *J. of Geotechnical Engrg., ASCE*, 115(8): 1148-1160.
Kmet, P. and D.E. Lindorff 1983. Use of Collection Lysimeters in Monitoring Sanitary Landfill Performance. Proceedings of the Characterization and Monitoring of the Vadose (Unsaturated) Zone, *National Water Works Assoc., 554-579.*
Morel-Seytoux, H.J. 1992a. Flux Determination at the Interface of Two Layers. *Colorado State University Hydrology Paper, March, 1992, Fort Collins, Colorado.*
Morel-Seytoux, H.J. 1992b. The Capillary Barrier Effect at the Interface of Two Soil Layers with Some Contrast in Properties, Second Report. *Hydrowar Reports Division,* Hydrology Days Pub., Fort Collins, Colorado.
Reades, D.W., L.R. Lahti, R.M. Quigley, and A. Bacopoulos 1990. Detailed Case History of Clay Liner Performance. *Waste Containment System: Construction, Regulation, and Performance*. R. Bonaparte Ed., Geotechnical Special Publication, No. 26, Nov. 1990.
Reades, D.W., R.J. Poland, and G. Kelly 1987. Discussion. *J. Geotechnical Engrg.,* ASCE, 113(7): 809-813.
Shackelford, C.D., C.-K. Chang, and T.-F. Chiu 1994. The Capillary Barrier Effect in Unsaturated Flow through Soil Barrier. *First International Congress on Environmental Geotechnics, July 10-15, 1994, Edmonton, Alberta, Canada*, Bi-Tech Publ. Ltd., Vancouver, B.C., 789-793

Tailings & Mine Waste'95 © 1995 Balkema, Rotterdam, ISBN 90 5410 526 7

Abandoned uranium mine reclamation – A case study

Edward T.Cotter

RUST Geotech Inc.*, US Department of Energy, Grand Junction Projects Office, Colo., USA

ABSTRACT: The U.S. Department of Energy Grand Junction Projects Office has developed a process for the reclamation of abandoned uranium mines located on lands under its administrative jurisdiction. This process allows for mine sites to be reclaimed in a cost-effective manner while maintaining a practicable approach to the presence and cleanup of naturally occurring radioactive materials resulting from past mining activities.

1 INTRODUCTION

The U.S. Department of Energy (DOE) Grand Junction Projects Office (GJPO) currently administers 43 uranium lease tracts located in southwestern Colorado (38), southeastern Utah (4), and northern New Mexico (1) consisting of approximately 25,000 acres of land (see Figure 1). These lands were withdrawn from the public domain from 1948 to 1954 by the U.S. Atomic Energy Commission, predecessor agency to the DOE, to develop a source of domestic uranium ores for defense purposes. Subsequent to withdrawal, these lands were included in the Uranium Leasing Program. As this program ended in 1962, all of the mines were abandoned and many of the mine portals were temporarily closed. However, little else was done to reclaim the land with environmental disturbances resulting from these mining activities. DOE–GJPO has adopted a policy for the reclamation of these areas. DOE–GJPO recognizes that, in addition to the abandoned uranium mines located within the boundaries of the lease tracts, many such mines exist on lands controlled by the U.S. Department of Interior Bureau of Land Management (BLM), the U.S. Department of Agriculture Forest Service (USFS), and other Federal and State agencies.

Abandoned uranium mines have naturally occurring radioactive materials (NORM) associated with them in the form of low-grade or residual ore materials that were not economic to transport and mill. In addition, outcrops of uranium-bearing formations often exist at the mine sites and represent the initial

---

*   Work supported by the U.S. Department of Energy, Office of Environmental Management, under Contract No. DE–AC04–86ID12584.

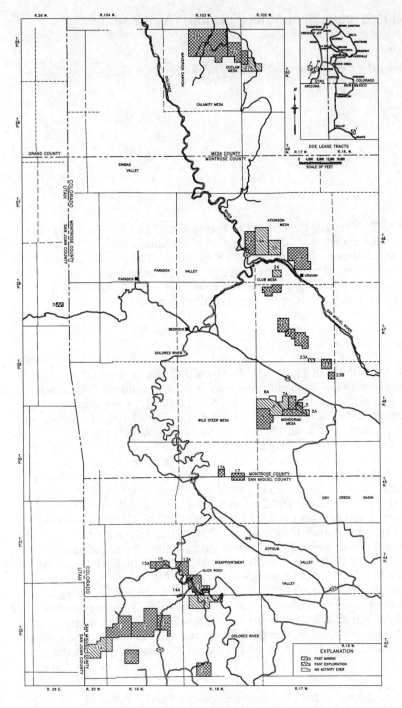

Figure 1. Location of U.S. Department of Energy uranium lease tracts.

discoveries that led to the mines' development. These radioactive materials are a potential source of radiation exposure to the general public and the environment. However, no regulations currently exist for the radiological cleanup of these abandoned uranium mines. The U.S. Environmental Protection Agency is in the process of proposing regulations that may address the presence of NORM at inactive uranium mines. DOE acknowledges that it is often cost prohibitive, if not entirely impossible, to eliminate these NORM sources from the environment.

2 RECLAMATION PROCESS

DOE has developed a process for the reclamation of abandoned uranium mines located on lands under its administrative jurisdiction. DOE's primary objectives in the reclamation of abandoned uranium mine sites are fourfold:
- Eliminate all physical safety hazards that have resulted from previous mining activities, including all mine openings and portals and all surface depressions that have severe vertical drop-offs of greater than 2 to 3 feet.
- Recontour the areas of disturbance to blend in with the natural topography surrounding the site, while at the same time, allowing as much vegetation regrowth to survive as possible.
- Eliminate or minimize the potential for mining-associated radiological materials to erode and migrate into natural drainages.
- Redirect all storm water away from the immediate areas where mine openings have been closed to eliminate the possibility of water flow and erosion into the mine workings and contain and control all storm water that contacts the site.
- Decrease the potential for the general public's exposure to radiological materials as practicable.

3 ELIMINATING PHYSICAL SAFETY HAZARDS

All mine openings/portals (shafts, adits, inclines, and vent holes) will be permanently closed by methods approved by the BLM. These methods include backfilling with rock (waste-dump) materials remaining at the mine site and in some instances, using man-made bulkheads. Selective placement of material within the mine openings/portals according to size is used to enhance slope stability and preclude future sloughing of the materials after closure is complete. After closure, all depressions surrounding or immediately adjacent to the mine openings/portals will be backfilled with additional waste-dump materials located at the site and then slightly mounded to preclude future subsidence.

4 RECONTOURING AREAS OF DISTURBANCE

After the mine openings/portals have been successfully and permanently closed, all areas of disturbance will be recontoured to blend in with the natural topography, as practicable. This includes

• Reducing the slopes of the remaining waste-dump materials to less than a 3-to-1 ratio while providing a basin effect on the top of the waste-dump material to collect and contain all storm water that contacts the site.

• Removing waste-dump materials from the existing, natural drainages, as practicable, and eliminating the potential for these materials to enter the drainages in the future.

• Recontouring the immediate areas to provide an undulating surface that closely represents the natural surrounding topography.

Any area of disturbance that has successfully reclaimed itself during the past several decades will not be redisturbed unless physical hazards exist that must be addressed. In either case, as much of the existing vegetation regrowth as practicable will be left during the reclamation efforts. Any topsoil that previously was stockpiled or is immediately available for use will be spread across the disturbed areas to promote revegetation. If topsoil has not been previously stockpiled or is not immediately available within the disturbed area, it will not be mined at another location and hauled in—a second area of disturbance will not be created just to enhance the reclamation efforts or success of the initial area. All disturbed areas will then be reseeded with a mixture of native grasses and shrubs developed for the specific climatic conditions and approved by the BLM. Silt fences or other similar devices will be installed in or across all drainages located immediately below the mine sites as an additional means of controlling the potential transport of sediments from the site by storm water.

5 REDIRECTING STORM WATER

As previously stated, several efforts will be undertaken to control the effects that storm water has on the abandoned mine site. First, all mine entrance locations will be backfilled sufficiently to provide positive drainage away from the entrance area. Secondly, the waste-dump materials will be recontoured in such a manner that a basin is created on top of the dump material to collect and contain all storm water that comes in contact with it. In addition, the basin will be constructed in such a way that mini-basins are created within the major basin feature to promote individual collection points that hopefully will enhance the revegetation efforts. The storm water will then be allowed to evaporate from and/or percolate through the basin. Next, all major drainage features leading onto the mine site will be diverted around the mine site or stabilized in place to reduce the overall effects that major storm water events will have on the reclamation efforts and the abandoned mine site in general. Finally, silt fences or other suitable devices will be placed within and across the drainages leaving the mine site to preclude the transport of sediments (possibly containing NORM) from the mine site to locations downstream.

6 REDUCING POTENTIAL EXPOSURE TO RADIOACTIVE MATERIALS

All abandoned uranium mines have residual radioactive material (NORM) associated with them. This material occurs as low-grade material that is not economic to transport and mill or as residual ore stockpiles. Outcrops of uranium-bearing formations often exist at the mine sites and represent the initial ore deposit that led to the mine's development. These radioactive materials are a potential source of radiation exposure to the general public and the environment. It is DOE's policy to minimize the potential exposure to the general public resulting from mining-related radioactive materials. However, it is often cost prohibitive, if not entirely impossible, to eliminate these NORM sources from the environment. Consequently, during reclamation activities on the DOE lease tracts, measures will be taken to limit the potential for exposure to NORM:

• During the preliminary reclamation activities, a cursory radiological scan of the entire mine site will be conducted to identify those areas of greatest radiological concern.

• All materials of greatest radiological concern will be used in the initial backfilling of the existing mine openings/portals so that the most radioactive material is placed or buried at depth below ground, leaving less radioactive material exposed at the ground surface.

• Additional scans will be conducted during the entire reclamation process to continually locate other areas of radiological concern for control or burial. After final recontouring is completed, any remaining areas of radiological concern will be further limited by covering the disturbed areas with any immediately available topsoil.

DOE has recently completed reclamation activities on the first phase of abandoned mines to be reclaimed by this process. These activities included work at multiple shallow subsurface mine sites (less than 100 feet below ground) and shallow surface mine sites (less than 20 feet deep) on two lease tracts (C–SR–16 and C–SR–16A) located in the Slick Rock area of southwestern Colorado.

Lease tracts C–SR–16 and C–SR–16A are located at the top of Slick Rock hill approximately 5 miles south of the community of Slick Rock, Colorado, in Sections 10, 11, 14, 15, and 16, Township 43 North, Range 19 West, New Mexico Principle Meridian. Both tracts are accessed by Colorado State Highway 141, which traverses lease tract C–SR–16A in a north–south direction. Several county roads also access the tracts from different points. The mesa-top topography is relatively flat but is broken up by the canyon features of the Dolores River and associated drainages. The area is covered primarily with stands of pinyon pine and Utah juniper intermixed with sagebrush parks. The region has an arid climate with normal precipitation between 12 and 16 inches per year. Land use in this area historically has been mining and grazing, although some dryland agricultural crops are grown nearby. Hunting is the primary recreational use of these lands. The mine sites within the tracts are located at an elevation of 7,100 to 7,200 feet above mean sea level. The uranium-bearing formation of interest in the Slick Rock area is the Salt Wash Member of the Morrison Formation, which lies at or very near the ground surface within the lease tracts. By the very nature of the geologic, hydrologic, and climatic conditions of the region, none of the mines associated with these lease tracts encountered groundwater, and consequently, groundwater was not an issue in the reclamation of these mines.

The mine sites reclaimed under this project were located in eight different areas; three areas located on lease tract C–SR–16 are designated by their associated claim names (Michael Bray, Ann, and Nola-Z), and five areas located on lease tract C–SR–16A are designated by their geographic location within the lease tract (central, north, northeast, northwest, and west).

The Michael Bray and Ann mine sites on lease tract C–SR–16 were very similar in that the ore horizon/zone of mineralization was entirely below grade and was accessed by shallow inclines. These mines had mine-specific waste dumps associated with them. Two of the Ann mines were somewhat unique in that they were equipped with ventilation shafts. Permanent closure of these mines, including the ventilation shafts, and the subsequent recontouring activities were relatively simple using conventional construction equipment (front-end loader, backhoe, and dozers). The mine sites were left slightly mounded to prevent storm water from draining into the underground workings.

The mine sites on lease tract C–SR–16A were all similar in that the ore horizon/zone of mineralization was very shallow, if not exposed at the surface, and was first accessed by various open pits. This led to extensive surface workings. Later, as the ore-bearing formation dipped further below grade, adits leading from the bottom of the open pits were used to mine the ore. The overburden material and associated mine waste were scattered throughout the different mine sites; some in large dumps, others in small piles from single ore carts. As before, permanent closure of these mines and the subsequent recontouring activities were relatively simple using conventional construction equipment. Mine sites with underground workings were left slightly mounded to prevent storm water from draining into the underground workings. The shallow open pit areas were left slightly bowled to contain and control storm water.

The Nola-Z mine site on lease tract C–SR–16 was distinctly unique from all of the other mine sites because it is located along the rim of Summit Canyon. The Nola-Z mine site comprised eight adits located approximately 25 feet below the canyon rim. During past mining operations, the mine waste materials were continuously deposited along the outer edge of the mine site access road that traversed the south slope of the canyon to the respective mines. The large boulders found above and adjacent to the mine portals along the canyon rim were considered to be extremely unstable and worker safety was a primary concern during the reclamation of this site. These unstable conditions prompted the use of explosives to close the portals. Vertical holes were drilled along the rim above the portals, and angle holes were drilled into the walls of the mine portals from an area directly in front of the respective mines. The holes were then loaded with explosives, which were detonated. This method proved very successful at seven of the eight adits. However, the operation had to be repeated to close the eighth and largest adit. The entire rim area was then dressed up with materials that were available along the surface of the rim. In addition, a berm was constructed above and along the rim to divert all storm water away from the reclaimed area. The mine waste dumps were left intact to act as a berm along the canyon wall to control storm-water runoff and limit further erosion of the material into the canyon.

During the project, a total of 38 mine portals (adits and inclines) and two ventilation shafts were permanently closed by methods approved by the BLM and the State of Colorado (Department of Natural Resources, Division of Minerals and Geology). In addition, approximately 50,000 cubic yards of material (mine waste, overburden, and topsoil) was backfilled into the depressions/open pits and recontoured to blend in with the surrounding topography. After the mine closures and recontouring were complete, all of the areas of disturbance (approximately 22 acres) were reseeded by mechanical and hand-held broadcast seeders with a seed mixture approved for use by the BLM.

7 CONCLUSION

The DOE administers 43 lease tracts (25,000 acres) that collectively contain a maximum of 200 abandoned and active uranium mine sites. This contrasts to millions of acres administered by the BLM, USFS, and other Federal and State agencies that contain several thousand abandoned and active uranium mine sites. The DOE feels that the aforementioned process provides a practical and cost-effective approach to abandoned uranium mine-site reclamation.

Tailings & Mine Waste'95 © 1995 Balkema, Rotterdam, ISBN 90 5410 526 7

Gypsum may delay acid production from waste rock

L. H. Filipek, C. S. E. Papp & J. Curry
United States Geological Survey, Lakewood, Colo., USA

T. J. VanWyngarden
ACZ Laboratories, Inc., Steamboat Springs, Colo., USA

T. R. Wildeman
Colorado School of Mines, Golden and Knight Piesold and Co., Denver, Colo., USA

ABSTRACT: From the results of two laboratory weathering studies, we postulate that the presence of gypsum delays onset of pyrite oxidation and acid production in waste rock from a porphyry copper/gold deposit located in an arid temperate environment. Acid-base accounting, sulfur speciation, and X-ray diffraction analyses of unweathered and weathered waste rock, and chemical analyses of leachate solutions were used to determine that pyrite and the little carbonate present in the rock remained relatively unaltered. Conversely, gypsum dissolution proceeded steadily throughout the 35-week duration of the tests. These results suggest that application of gypsum, perhaps as a brine solution, could be used to suppress acid generation in arid environments.

1 DEPOSIT GEOLOGY

Material for this study came from a porphyry copper/gold deposit in an arid region of the Andes that receives its major moisture during a few storms in the summer. Samples were waste rock obtained from drill holes, mainly in unweathered parts of the deposit. The presence of quartz, gypsum, mica, plagioclase, pyrite, chlorite, and potassium feldspar was determined by X-ray diffraction analysis. Very little carbonate was detected, with an average of about 0.1 percent carbon. The presence of gypsum was not anticipated, because gypsum does not typically occur in porphyry systems. However, anhydrite has been found in the central, potassic alteration zone of porphyry systems (Lowell and Gilbert, 1970). Thus, the gypsum could be a hydrated product of anhydrite. Alternatively, gypsum could have been deposited along fractures from evaporated saline groundwater. Gypsum commonly occurs as an evaporation product within soils in this type of climate.

2 WEATHERING STUDIES

Two weathering studies were made, each having 35 weekly weathering cycles. The weekly cycle for both sets of tests was similar: Waste rock was (1) exposed to dry air for three days, then (2) exposed to water-saturated air for three days, and finally (3) flooded with water on the seventh day at a one-to-one water-to-rock weight ratio to leach any reaction products. One test was conducted in an EPA simulated humidity cell (Sobek and others, 1978, Method 4.1.5); the other test was conducted in a 10-cm diameter, 30-cm long PVC column. In the humidity cell test, 250 g of waste rock was ground to <2.0 mm to produce a homogeneous sample. In the column test, a mix of particles to 5 cm nominal diameter was used to better represent the heterogeneous nature of a waste-rock pile. The two types of weathering studies were conducted in parallel to determine the effect of particle size and distribution on rate of onset of acid generation.

Acid-base accounting (Sobek and others, 1978) and a second recently developed sulfur speciation scheme (Curry and Papp, unpublished) were used to estimate the acid production potential (APP) of composites of the unweathered and weathered waste rock after 35 weeks of leaching (see Table 1). Acid-base accounting is a means of determining whether a rock can produce acid drainage. The APP is a function of the concentration of iron sulfides, especially pyrite, because such sulfides produce sulfuric acid when they oxidize. The Sobek acid-base accounting method is a series of extractions to separate sulfate (hydrochloric acid-extractable), pyrite (nitric acid-extractable), and organic (residual) sulfur, where it is typically assumed that the nitric acid-extractable sulfur is the only sulfur form that produces acid. This scheme was designed for analysis of sulfur in coals. In the present study, the Sobek sulfur-speciation scheme was modified in that the sum of the fractions designated "nitric acid-extractable sulfur" and "residual sulfur" was considered the acid-producing component of sulfur, labeled "sulfide sulfur," because little or no organic sulfur was expected in these igneous rocks. The Curry and Papp sulfur speciation method is designed specifically for non-coal rocks and uses a sodium pyrophosphate solution to extract the organic sulfur fraction. The two methods gave fairly similar results for the sulfur species (Table 1).

Table 1. Unweathered and Weathered Waste Rock: Sulfur forms, acid production potential (APP), acid neutralization potential (ANP), and net acid production potential (NAPP)

	Test[1]	Total S (%)	Sulfate S (%)	Sulfide S (%)	APP[2]	ANP[2]	CaCO$_3$[2]	NAPP[2]
Unweathered	EPA	7.3	3.4	3.9	123	24		99
	C P	8.1	3.8	4.2	132		8	108
Humidity Cell	EPA	5.7	1.9	3.8	119	20		99
	C P	6.2	1.8	4.4	138		7	118
Column	EPA	5.7	0.8	4.6	145	23		122
	C P	5.8	1.2	4.6	143		8	120

[1]Sulfur speciation was determined by two different test methods:
    EPA = Sobek and others (1978).
    C P = Curry and Papp (unpublished).

[2]In g of CaCO$_3$ equivalent per kg rock.

The acid neutralization potential (ANP) is typically mainly a function of the calcium and magnesium carbonate content. The acid-base accounting calculation assumes that the neutralizing component consists entirely of calcium carbonate. In calculating APP, a conversion factor of 31.25 is typically used to convert weight percent sulfide sulfur to the equivalent weight of calcium carbonate needed to neutralize the acid, in grams per 1,000 grams (g/kg). This conversion allows a direct comparison between APP and ANP values.

3 TESTING RESULTS

The waste rock had a net APP (NAPP) of about 100 to 120 g/kg of calcium carbonate equivalent, based on the difference between APP and ANP. After 35 weeks of simulated weathering, neither weathered composite was significantly different from the unweathered composite in terms of APP, ANP, or NAPP, which suggests that very little pyrite or carbonate were dissolved (Table 1).

The leachate from each weekly cycle of the weathering tests was analyzed for pH, conductivity, alkalinity, acidity, and content of calcium, magnesium, iron, and sulfate. The solutions from both sets of tests were essentially poorly buffered solutions of calcium sulfate with minor magnesium. During the 35-week time period, the pH of the column leachate remained about 4.5, while that of the humidity cell varied between 6.4 and 7.1. The cumulative leaching curves for calcium and sulfate in both tests were linear over the 35 weeks, whereas most of the magnesium leaching occurred over the first 15 to 20 weeks. Initially, the mole ratio of calcium to sulfate in the leachate from the humidity cell was about 0.8 and from the column, 0.6. The ratio gradually increased to the one-to-one molar ratio of gypsum in both tests. X-ray diffraction analyses of the weathered waste rock indicated a decrease in the ratio of gypsum to pyrite compared to the unweathered rock, confirming that the source of the leached sulfate was gypsum.

4 FUTURE WORK

Additional work is underway to investigate the mechanisms by which gypsum acts to delay onset of acid generation. Thin sections of the unweathered rock and grain mounts of pyrite from the weathered rock are being prepared for scanning electron microscope analysis. The goals of these investigations are to determine whether the gypsum coats the pyrite, armoring it from exposure to oxygen and whether the pyrite from the weathered rock shows any evidence of surface reaction.

REFERENCES

Lowell, J.D. & J.M. Guilbert. 1970. Lateral and vertical alteration-mineralization zoning in porphyry ore deposits, *Economic Geology* 65:373-408.
Sobek, A.A., Shuller, W.A., Freeman, J.R., & R.M. Smith. 1978. *Field and laboratory methods applicable to overburdens and minesoils.* EPA 600/2-78-054.

Geochemical modeling of waste rock leachate generated in precious metal mines

John P. Kaszuba, Wendy J. Harrison & Richard F. Wendlandt
Department of Geology and Geological Engineering, Colorado School of Mines, Golden, Colo.,
USA

ABSTRACT: Geochemical calculations using the reaction path model EQ3/6 examine aqueous equilibrium relationships among minerals and waters in waste rock typically generated by precious metals mines in the Western United States. Freshwater infiltration of waste rock may lead to pyrite oxidation and acid leachate production if significant amounts of sulfide minerals are present. Aluminosilicate minerals consume protons in a typical waste rock system at pyrite contents up to 4% by volume when the system is closed to a constant oxygen supply from the atmosphere. Sensitivity studies indicate that protons are produced at higher concentrations of pyrite (>4% by volume), in environments where oxygen is in infinite supply, and where relative oxidation rates exceed aluminosilicate reaction rates by 1 to 2 orders-of-magnitude. The bulk rock's ability to consume protons in the leachate is due to hydrolysis of aluminosilicate minerals. Calcite improves the system's ability to consume protons. In actual waste rock systems, pyrite contents between 0.3% and 1.4% by volume produce acid leachate since the relative rate of pyrite oxidation exceeds that of aluminosilicate hydrolysis by 10 to 100 times.

1 INTRODUCTION

The oxidation of pyrite is an important natural process that takes place in environments where the geosphere and the hydrosphere collide, producing acidity and dissolved metals in aqueous environments. Two geochemical reactions critical to this process are:

(1) $FeS_2 + 7/2\ O_2 + H_2O = 2\ SO_4^{--} + 2\ H^+ + Fe^{++}$, and

(2) $FeS_2 + 14\ Fe^{+++} + 8\ H_2O = 2\ SO_4^{--} + 16\ H^+ + 15\ Fe^{++}$

(Singer and Stumm 1970). Environments where these or similar geochemical reactions produce acid leachate, such as mine tailings, ore piles, mine shafts, pyritic shales, and coal dumps, have been the focus of intense scrutiny for a great number of communities, government bodies, industrial organizations, and research groups (Ahmad 1974; McWhorter *et al.* 1974; Ralston and Morilla 1974; U.S. Department of the Interior 1994).

Of particular interest, however, are the geologic environments where pyrite oxidation does not result in acid mine drainage. The protons (acidity) that we know are produced by pyrite oxidation are consumed by other processes within these environments. An understanding of the relationship between pyrite oxidation and these other processes may improve remediation strategies for acid mine drainage and management practices for mine waste. The most effective and widely-distributed proton-consuming reactions are those involved in mineral weathering, specifically the hydrolysis reactions of the aluminosilicate minerals.

In this study we investigate one such environment. We examine aqueous equilibrium relationships between waste rock and leachate typically generated by precious metal mines in the western United States. Waste rock is material removed from the mine in order to reach economically recoverable metal ores. It may be stockpiled in heaps or used as backfill in open pit mines. Freshwater infiltration may lead to pyrite oxidation and acid leachate production if the waste rock contains a significant amount of sulfide minerals.

2 METHODS

The interaction between waste rock and leachate may be thought of as a sequence of irreversible chemical reactions. These reactions drive the chemical evolution, or reaction path, of the system. As the reactions proceed, minerals and other solids precipitate or dissolve. The chemistry of the leachate changes in conjunction with the solid equilibria. Calculation of these mineral and aqueous equilibria provides an understanding of the mineral assemblages and water chemistry produced by the system.

The computer program EQ3/6 was used to make the calculations. EQ3/6 is a speciation-solubility and reaction path computer program that examines the chemical evolution of a water/rock system as reaction progress or time advances (Wolery 1983; Wolery and Daveler 1989).

The chemistry of fresh water that initially infiltrates the waste rock and the mineralogy of the rock are required as input for the calculations. The chemical analysis of the water used as input (Table 1) is modified from the average fresh water composition of Long and Angino (1977). In a mine waste environment, the source of the fresh water is generally surface water and/or groundwater. Surface water may infiltrate the waste rock through surface run-off and/or direct precipitation. This water is initially in equilibrium with atmospheric oxygen and carbon dioxide. Groundwater may infiltrate the waste rock by way of discharge from any suitable aquifer. Confined aquifers or deeper groundwater systems, however, yield water that generally contains little dissolved oxygen and carbon dioxide in concentrations greater than those observed in surface waters (Freeze and Cherry 1979; Driscoll 1987; Fetter 1988). Geochemical calculations were made using waters derived from each of these two sources. Oxygen and carbon dioxide values of 8.5 milligrams per liter (mg/l) and 0.5 mg/l for surface water, and 0 mg/l and 14.7 mg/l for groundwater, respectively, were used. The calculations assume the hydrology of both sources allows the respective waters to migrate into and accumulate within the waste rock.

The mineralogy of a waste rock heap depends on the regional geology as well as the ore body of interest. For our calculations we selected an occurrence of precious metals ores that is representative of many deposits in the Western United States (Schafer *et al*. 1988; Davis and Streufert 1990; Raines *et al*. 1991) and made several simplifying assumptions. The host rock for mineralization is a granitic gneiss which is overlain by a veneer of alluvium. Pyrite is the only sulfide mineral that is present in the mineralized granitic gneiss disposed as waste rock. The mineralogy of the alluvium is that of an average sandstone (Blatt *et al*. 1980) containing no carbonate and 2 vol% clay minerals. Waste rock is disposed as a mix of alluvium and mineralized gneiss. The mineralogy of the waste rock that is constructed with these assumptions is presented in Table 2. The modal data listed in Table 2 were converted to molar data for the calculations using the molar volumes for each mineral. The conversion was made assuming a porosity of 30% and a normalized fluid volume of 1 liter. Mixed unconsolidated sand and gravel have a porosity that ranges from 20% to 35% (Fetter 1988). Minerals occurring as solid solutions were input into EQ3/6 as subequal amounts of discrete phases of pure end-member components. We also assumed that reaction rates for the minerals are proportional to their modal abundances so that each mineral dissolves at a rate proportional to its abundance in the rock.

24

Table 1. Fresh water chemistry (mg/l). This water has a pH of 6. Modified from Long and Angino (1977).

Ca	15	F	0.005
Mg	4.1	HPO_4	0.06
Na	6.3	Fe	0.002
K	2.4	Cu	7×10^{-3}
Cl	7.8	Zn	5×10^{-3}
SO_4	96	Pb	5×10^{-5}
SiO_2	9	Al	0.002

Table 2. Mineralogy of waste rock. Pure end-member components are also listed for minerals occurring as solid solutions. Geochemical calculations used subequal amounts of end-member components in the solid solutions in order to simplify the calculations. Volume % and number of moles are related by the molar volume for each mineral. Volume % and weight % are related by the by the molar volume and molecular weight for each mineral. Molar volumes and molecular weights are tabulated in Robie *et al.* (1979), Hemingway *et al.* (1982), and in the EQ3/6 data base. For clay minerals, these conversions are not reliable if microporosity is present in the clay.

Mineral	Volume %	Weight %	# of Moles
Quartz	27	34	39.3
K feldspar	21	26	6.4
Plagioclase	0.4		
(Albite and		0.4	0.1
Anorthite)		0.4	0.1
Clay	2		
(Kaolinite and		1.1	0.3
ideal Illite)		4.5	0.2
Muscovite	5.5		
(Muscovite and		5.2	0.9
Paragonite)		2.2	0.4
Biotite	4		
(Phlogopite and		3.0	0.5
Annite)		3.0	0.4
Amphibole	3		
(Tremolite and		2.4	0.2
Pargasite)		2.4	0.2
Chlorite	3		
(Clinochlore and		1.6	0.2
14A-Daphnite)		2.2	0.2
Hematite	2	6.0	2.6
Magnetite	2	5.0	1.5
Pyrite	0.3	0.7	0.4
Porosity	30	---	---
TOTAL	100.2	100.1	

Other constraints on the calculations, in addition to water chemistry and waste rock mineralogy, include a system temperature of 25°C and a total pressure of 1 bar. We assume the following conceptual design:

1. The waste rock/water mixture is a completely-mixed, closed system at steady state. In this system, a single pore volume of fresh water progressively reacts with the waste rock.

2. Wall rocks do not interact with the waste rock/water mixture.

3. Mineral reactions are dominated by equilibrium, not kinetic, processes.

To identify and understand the variables critical to acid leachate production resulting from surface water infiltration, we systematically varied:

1. The amount of pyrite present in the waste rock. The initial calculation used 0.3 vol% pyrite. Subsequent calculations evaluate pyrite contents ranging from 1.4% to 70% by volume. In order to maintain the 30% porosity and fluid volume of 1 liter, modal abundances of the other minerals in the waste rock were proportionately reduced.

2. Availability and effect of ferric iron as an oxidant. Most of the iron in the infiltrating fluid is present as ferrous iron. All of this iron was converted to the ferric state for subsequent calculations.

3. Oxygen content of infiltrating water. The oxygen content of the infiltrating fluid was initially specified as 8.5 mg/l, as determined by equilibrium with atmospheric oxygen. In the closed system, this value decreases as oxygen is consumed by reactions (1) and (2). To evaluate the effect of opening the system to the atmosphere, which is an infinite oxygen reservoir, the oxygen content of the leachate was specified to remain 8.5 mg/l for all portions of reaction progress in subsequent calculations.

4. Presence of carbonate minerals. No carbonate minerals were present in the initial calculations. However, carbonate minerals are known to consume protons. For example:

(3) $CaCO_3 + H^+ = Ca^{++} + HCO_3^-$

In order to quantify this geochemical process, 1% and 5% calcite (by volume) were added to the system for subsequent calculations. Modal abundances of the other minerals present in the waste rock were proportionately reduced to maintain the 30% porosity and normalized fluid volume of 1 liter.

5. Contribution of principal mineral groups to the system's buffering capacity. Each of the minerals which comprises the waste rock weathers according to a different reaction or set of reactions. For example:

(4) $2\ KAlSi_3O_8 + 2\ H^+ + 9\ H_2O = Al_2Si_2O_5(OH)_4 + 2\ K^+ + 4H_4SiO_4^\circ$

In this reaction, each mole of K-feldspar consumes one mole of protons. However, alternate weathering reactions for K-feldspar, as well as reactions for other alumino-silicate minerals, may not require this one-to-one ratio. In order to evaluate the net contribution of each of the principal mineral groups to the bulk rock's ability to buffer acidity, calculations were conducted for several different mineral assemblages. For each of these calculations, the minerals in each assemblage are present in the same molar amounts as specified in the original calculation (Table 2).

6. Rates of pyrite oxidation reactions relative to rates of aluminosilicate weathering reactions. Reaction rates for the minerals present in the waste rock are specified as proportional to their modal abundances. Reaction rates for pyrite were increased by a factor of 10 and 100 in subsequent calculations.

3 RESULTS AND DISCUSSION

The results of two initial calculations examining surface water and groundwater infiltration with 0.3 vol% pyrite in the waste rock are presented in Table 3. No acid

26

Table 3. Results of geochemical calculations for infiltration of surface water and groundwater into the waste rock.

Amount of pyrite		Minimum pH after fresh water infiltration	
Volume %	Modal ratio[*]	Surface water	Groundwater
0.3[**]	232	6.5	6.2
1.4	49	6.3	6.0
2.8	24	5.2	5.8
4	16.5	4.5	4.4
5	13	4.2	4.3
10	6	4.1	4.0
20	2.5	4.0	3.7
70	0	3.9	3.0

[*]Volume ratio of aluminosilicate minerals to pyrite
[**]A pyrite content of 0.3 vol% is the sulfide content of the original calculation

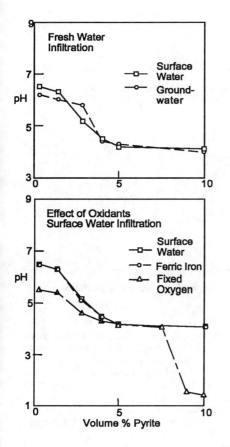

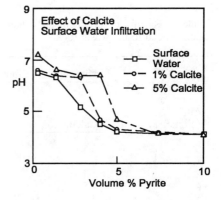

Figure 1. Relationships between amount of pyrite and pH generated in leachate. Data plotted for infiltration of surface water and ground water, effect of calcite, and effect of the oxidants ferric iron and oxygen.

27

Table 4. Results of geochemical calculations using surface water infiltration. Oxygen content is fixed, ferric iron replaces ferrous iron.

Amount of pyrite (Volume %)	Minimum pH after surface water infiltration		
	Initial Calculation	Fixed Oxygen Content	Ferric Iron
0.3*	6.5	5.5	6.5
1.4	6.3	5.4	6.3
2.8	5.2	4.6	5.1
4	4.5	4.3	4.5
5	4.2	4.2	4.2
7.5	---	4.1	---
9	---	1.5	---
10	4.1	1.4	4.1
20	4.0	---	4.0
70	3.9	0.2	3.8

*A pyrite content of 0.3 vol% is the sulfide content of the original calculation

Table 5. Results of geochemical calculations using surface water infiltration with calcite added to the waste rock.

Amount of pyrite (Volume %)	Minimum pH after surface water infiltration		
	Initial Calculation	1% Calcite	5% Calcite
0.3*	6.5	6.6	7.2
1.4	6.3	6.4	6.6
2.8	5.2	6.3	6.4
4	4.5	4.7	6.4
5	4.2	4.3	4.7
7.5	---	4.1	4.1
10	4.1	4.1	4.1
20	4.0	4.0	4.0
70	3.9	3.8	3.8

*A pyrite content of 0.3 vol% is the sulfide content of the original calculation

leachate was produced in either calculation. Instead, the pH of the leachate increased for both surface water (pH = 6.5) and groundwater (pH = 6.2) infiltration relative to the pH of the water (pH = 6) that initially infiltrated the waste rock. All of the minerals present in the waste rock, except quartz and kaolinite, dissolve as the rock and the water react with one another. Quartz and kaolinite do not dissolve because both surface water and groundwater are saturated with respect to these two minerals.

Subsequent calculations that evaluate acid generation for a range of pyrite contents are summarized in Table 3 and Figure 1. In general the pH of the leachate decreases

Table 6. Results of geochemical calculations for interaction of surface water with selected mineral assemblages that occur in the waste rock. The minimum pH generated for each of the listed assemblages containing the specified amount of pyrite is tabulated. These data are discussed in the text to evaluate each principal mineral group's contribution to the system's ability to consume protons.

Assemblage[1]	Minimum pH at pyrite contents (volume %) of:			
	0.3	1.4	2.8	4
(1) pyrite	3.8	3.8	---	---
(2) pyrite + quartz	3.8	3.8	---	---
(3) pyrite + K-feldspar	6.7	4.5	4.2	---
(4) pyrite + K-feldspar + quartz[2]	6.7	4.5	4.2	---
(5) pyrite + plagioclase	4.0	3.8	---	---
(6) pyrite + clay	4.0	4.0	---	---
(7) pyrite + mica	6.0	4.2	---	---
(8) pyrite + amphibole	6.0	4.2	---	---
(9) pyrite + chlorite	6.0	3.9	---	---
(10) pyrite + hematite + magnetite	3.9	3.8	---	---
(11) pyrite + 1% calcite	5.8	3.9	---	---
(12) pyrite + mica + amphibole + chlorite	6.0	6.0	4.4	---
(13) pyrite + plagioclase + mica + amphibole + chlorite	6.0	6.0	4.5	---
(14) pyrite + quartz + K-feldspar + plagioclase + mica + amphibole + chlorite + hematite + magnetite	6.0	6.0	6.0	4.6

[1]These assemblages have not been normalized to 1 liter of water; volume of fluid ranges from approximately 2.5 to 3.3 liters
[2]This assemblage normalized to 1 liter of water by increasing amount of quartz that is present to reduce porosity to 30%

by approximately 1 pH unit with each ten-fold increase of pyrite down to a minimum pH of 3.9. Assuming that a pH of 4.5 or less defines acid leachate conditions, our calculations indicate that the interaction between surface water and waste rock generates acid leachate once a pyrite content of 4 vol% is present in the waste rock. Below 4% pyrite, the leachate that is produced has a pH ranging from 5.2 to 6.5. The interaction between groundwater and waste rock generates acid leachate once a pyrite content of 3.9 vol% is present. The most acidic leachate (pH = 3.9) is generated when the waste rock is composed wholly of pyrite.

The calculations that evaluate the effect of ferric iron as an oxidant are presented in Table 4 and Figure 1. The leachate that is generated when all iron is present as ferric iron exhibits a pH almost identical to the pH of leachate produced when all iron is present as ferrous iron. In all of the calculations, ferric iron persisted throughout the chemical evolution of the system and was not reduced to ferrous iron. However, since the total iron content of the infiltrating fluid is rather low (Table 1), the supply of oxidant in the form of iron may be so small that pyrite oxidation according to reaction (2) is insignificant. In systems containing a greater amount of iron, reaction (2) may be more important.

The effect of establishing a fixed oxygen content in the leachate is summarized in Table 4 and illustrated in Figure 1. At pyrite contents of 1.4 vol% and less, the pH of the leachate is approximately 1 pH unit lower than the pH of leachate generated without fixing the oxygen content. Leachate in equilibrium with 4% and 5% pyrite (by volume) has a pH that is very similar for both calculations. Pyrite contents of 10 vol% and greater generate leachate that is approximately 2.5 pH units lower than leachate generated without fixing the oxygen content. When the system is composed wholly of pyrite, a pH of 0.2 is generated. Assuming that a pH of 4.5 or less defines acid leachate conditions, our calculations indicate that the interaction between surface water saturated with oxygen and waste rock generates acid leachate at a pyrite content of approximately 2.8 vol%.

Clearly, opening the water/rock system to an unlimited oxygen reservoir profoundly effects acid production. The system is only able to buffer acidity below a pyrite concentration of 2.8 vol% (Figure 1), a value that is one-third smaller than when the system is closed to oxygen. An especially acidic leachate with pH less than 2 is produced when the pyrite contents approach 10 vol%. These data suggest that oxygen content in the water, and consequently oxygen supply to the water/rock matrix, is a critical factor in the generation of acid mine drainage. The supply of oxygen is more critical to the generation of acid leachate than the supply of ferric iron as an oxidant.

The calculations that quantify the addition of calcite to the waste rock are summarized in Table 5 and illustrated in Figure 1. Addition of 1 vol% calcite maintains a pH in the leachate greater than 6 up to pyrite contents of 3 vol%. Acid leachate is produced once pyrite levels reach 4.5 vol% (Figure 1). Addition of 5 vol% calcite maintains a pH in the leachate greater than 6 up to pyrite contents of 4.25 vol%. Acid leachate is produced once pyrite levels reach approximately 6 vol%. These data suggest that the acidity produced by pyrite oxidation which is not consumed by the bulk rock is consumed by the weathering of calcite, perhaps according to reaction (3). These data (Figure 1) also suggest that calcite is able to buffer this surplus acidity up to a pyrite:calcite ratio of 0.4 to 0.5 by volume. The presence of calcite in the waste rock clearly improves the bulk rock's ability to consume protons.

The evaluation of each principal mineral group's contribution to the system's buffering capacity is summarized in Table 6. In general, K-feldspar, mica, amphibole, chlorite, and small amounts of calcite (less than 1 vol%) are each good buffering agents when paired with less than 1.4 vol% pyrite. Quartz, plagioclase, clay, hematite, and magnetite do not consume protons in this system. Assemblages which include several different types of aluminosilicate minerals (e.g., assemblages #12, #13, and #14 listed in Table 6) consume protons more effectively than assemblages that pair pyrite with only one type of aluminosilicate (e.g., assemblages #3 and #11 in Table 6). With the exception of quartz and kaolinite, each of the minerals in the waste rock is under-saturated with respect to the leachate. These minerals weather by interaction with the leachate. Weathering of the aluminosilicate minerals consumes protons according to geochemical reactions such as (4). Illite and plagioclase in this system probably do not buffer well because they are present in such low amounts.

Evidently, the bulk rock's ability to buffer acid production in the leachate is due to weathering of aluminosilicate minerals. Strömberg and Banwart (1994) modeled mine waste from a copper mine in northern Sweden. Their calculations indicate that, in decreasing order of importance, biotite, anorthite, albite, and K-feldspar weathering are critical processes in the consumption of protons within waste rock at the copper mine. The waste rock of Strömberg and Banwart (1994) contains similar amounts of quartz, K-feldspar, and pyrite but twice the mica and 50 times the plagioclase content of our system. Watershed studies suggest that weathering of plagioclase (Kirkwood and

30

Table 7. Results of geochemical calculations for surface water infiltration into waste rock comparing rate of pyrite oxidation reactions relative to rates of aluminosilicate weathering reactions.

	Minimum pH after surface water infiltration		
Amount of pyrite (Volume %)	Initial Calculation	Pyrite reaction rate increased 10X	Pyrite reaction rate increased 100x
0.3*	6.5	6.3	4.0
1.0	---	4.1	---
1.4	6.3	4.0	4.0
2.8	5.2	4.0	---
4	4.5	4.0	4.0
5	4.2	4.0	---
10	4.1	4.0	4.0
20	4.0	4.0	4.0

*A pyrite content of 0.3 vol% is the sulfide content of the original calculation

Nesbitt 1991), ferromagnesian silicates (Miller and Drever 1977), or both (Velbel 1985; Giovanoli et al. 1988) are important to the consumption of protons.

Calculations that evaluate rates of pyrite oxidation reactions relative to rates of aluminosilicate weathering reactions are summarized in Table 7. A ten-fold increase in pyrite oxidation relative to aluminosilicate weathering produces acid leachate at all calculated pyrite concentrations greater than 0.3 vol%. A 100-fold increase produces acid leachate at all pyrite concentrations.

Strömberg and Banwart (1994) determined characteristic residence times for minerals in their waste rock based on results of their calculations and on published stoichiometries and rate laws for important geochemical reactions (Helgeson et al. 1984; Wiersma and Rimstidt 1984; Chou and Wollast 1985; McKibben and Barnes 1986; Amrhein and Suarez 1988; Knauss and Wolery 1989; and Williamson 1992). Strömberg and Banwart's (1994) results suggest that pyrite weathers at a relative rate that is 4, 7, 14, 57, and 4,300 times faster than the rate for anorthite, biotite, albite, K-feldspar, and muscovite weathering, respectively. Our relative reaction rate calculations are within the same order-of-magnitude as these residence time calculations. Consequently, since pyrite oxidation will proceed at a rate 10 to 100 times faster than aluminosilicate weathering, pyrite concentrations of 0.3% to 1.4% by volume will produce acid leachate in natural waste rock systems.

4 CONCLUSIONS

Aluminosilicate minerals consume acidity in a typical waste rock system at pyrite contents up to 4 vol% when the system is closed to a constant oxygen supply from the atmosphere. Acidity is produced at higher concentrations of pyrite (>4% by volume), in environments where oxygen is in infinite supply, and where relative oxidation rates exceed aluminosilicate reaction rates by 1 to 2 orders-of-magnitude.

Oxygen supply a critical control of acid leachate production. When open to the atmosphere, the waste rock system is only able to buffer acidity below pyrite concentrations of 2.8 vol%, a value that is one-third smaller than when the system is closed to oxygen. An exceptionally acidic leachate with pH less than 2 is produced when the pyrite contents approach 10 vol% and the system is open to oxygen. The occurrence of all iron as ferric iron does not increase the system's ability to generate acid leachate.

Calcite improves the system's ability to buffer acidity. Addition of 1 vol% calcite to the waste rock increases the amount of pyrite it is able to buffer by 13 vol%. Addition of 5 vol% calcite increases the level of pyrite by 50 vol%. Calcite is able to buffer acidity not consumed by aluminosilicate minerals in the waste rock up to a pyrite:calcite ratio of 0.4 to 0.5.

In the absence of calcite, the bulk rock's ability to buffer acid production in the leachate is due to weathering of aluminosilicate minerals. K-feldspar, mica, amphibole, and chlorite contribute significantly to the buffering capacity of this system. Quartz, plagioclase, clay, hematite, and magnetite do not consume acidity in this system. Illite and plagioclase probably do not buffer well in this system because they are present in such low amounts.

Acid leachate is produced at all concentrations of pyrite when pyrite reaction rates are increased 100 fold relative to aluminosilicate weathering rates. When pyrite reaction rates are increased 10 fold, pyrite concentrations as little as 0.3 vol% will produce acid leachate. Characteristic residence times determined for minerals occurring in waste rock (Strömberg and Banwart 1994) suggest that pyrite weathers approximately 5 to 60 times faster than the aluminosilicate minerals anorthite, biotite, albite, and K-feldspar. Consequently, pyrite concentrations as small as 0.3% to 1.4% by volume will produce acid leachate in natural waste rock systems.

These conclusions possess important implications for disposal of mining waste rock and remediation of certain acid mine drainage problems despite the simplifying assumptions of ignoring sorption, organic activity, and specific kinetic rate laws and creating a system that achieves chemical equilibrium within an ideal hydrologic environment. In particular, the bulk waste rock possesses an intrinsic ability to consume acidity, even in the absence of carbonate minerals. The supply of oxygen to the water/rock matrix is critical to the production of acid leachate. Laboratory experiments and bench-scale tests that evaluate the use of limestone and other materials to prevent or remediate acid leachate (e.g. Day 1994; Rose and Daub 1994) must take into account these findings before the actual disposal or remediation activities that are based on the experiments begin.

5 ACKNOWLEDGEMENTS

Support from NSF grant EAR-9316197 is gratefully acknowledged.

REFERENCES

Ahmad, M.U. 1974. Coal mining and its effect on water quality. *American Water Resources Association Proceedings* 18:138-148.

Amrhein, C. & D.L. Suarez. 1988. The use of a surface complexation model to describe the kinetics of ligand-promoted dissolution of anorthite. *Geochimica Cosmochimica Acta* 52:2785-2793.

Blatt, H., G. Middleton & R. Murray. 1980. *Origin of Sedimentary Rocks.* 2nd Edition. Prentice-Hall, Inc.

Chou, L. & R. Wollast. 1985. Steady-state kinetics and dissolution mechanisms of albite. *American Journal of Science* 285:963-993.

Davis, M.W. & R.K. Streufert. 1990. Gold occurrences of Colorado. *Colorado Geological Survey Resource Series* 28.

Day, S.J. 1994. Evaluation of acid generating rock and acid consuming rock mixing to prevent acid mine drainage. Proceedings of the International Land Reclamation and Mine Drainage Conference and 3rd International Conference on the Abatement of Acidic Drainage. *Bureau of Mines Special Publication* SP 06B-94:77-86.

Driscoll, F.G. 1987. *Groundwater and Wells*. Johnson Division. St. Paul, Minnesota.

Fetter, C.W. 1988. *Applied Hydrology*. 2nd edition. MacMillan Publishing Co. New York.

Freeze, R.A. & J.A. Cherry. 1979. *Groundwater*. Prentice Hall. New Jersey.

Giovanoli, R., J.L. Schnoor, L. Sigg, W. Stumm & J. Zobrist. 1988. Chemical weathering of crystalline rocks in the catchment area of acidic Ticino Lakes, Switzerland. *Clays and Clay Mineralogy* 36:521-529.

Helgeson, H.C., W.M. Murphy & P. Aagaard. 1984. Thermodynamic and kinetic constraints on reaction rates among minerals and aqueous solutions. II. Rate constants, effective surface area, and the hydrolysis of feldspar. *Geochimica Cosmochimica Acta* 48:2405-2432.

Hemingway, B.S., J.L. Haas, Jr. & G.R. Robinson, Jr. 1982. Thermodynamic properties of selected minerals in the system Al_2O_3-CaO-SiO_2-H_2O at 298.15 K and 1 bar (10^5 pascals) pressure and at higher temperatures. *Geological Survey Bulletin* 1544.

Kirkwood, D.E. & H.W. Nesbitt. 1991. Formation and evolution of soils from an acidified watershed: Plastic Lake, Ontario, Canada. *Geochimica Cosmochimica Acta* 55:1295-1308.

Knauss, K.G. & T.J. Wolery. 1989. Muscovite dissolution kinetics as a function of pH and time at 70°C. *Geochimica Cosmochimica Acta* 53:1493-1501.

Long, D.T. & E.E. Angino. 1977. Chemical speciation of Cd, Cu, Pb, and Zn in mixed freshwater, seawater, and brine solutions. *Geochimica Cosmochimica Acta* 41:1183-1191.

McKibben, M.A. & H.L. Barnes. 1986. Oxidation of pyrite in low temperature acidic solutions: Rate laws and surface textures. *Geochimica Cosmochimica Acta* 50:1509-1520.

McWhorter, D.B., R.K. Skogerboe & G.V. Skogerboe. 1974. Potential of mine and mill spoils for water quality degradation. *American Water Resources Association Proceedings* 18:123-137.

Miller, W.R. & J.I. Drever. 1977. Chemical weathering and related controls on surface water chemistry in the Absaroka Mountains, Wyoming. *Geochimica Cosmochimica Acta* 41:1693-1702.

Raines, G.L., R.E. Lisle, R.W. Schafer & W.H. Wilkinson, eds. 1991. Geology and ore deposits of the Great Basin. *Symposium Proceedings, Geological Society of Nevada*. Reno, Nevada.

Ralston, D.R. & A.G. Morilla. 1974. Ground water movement through an abandoned tailings pile. *American Water Resources Association Proceedings* 18:174-183.

Robie, R.A., B.S. Hemingway & J.R. Fisher. 1979. Thermodynamic properties of minerals and related substances at 298.15 K and 1 bar (10^5 pascals) pressure and at higher temperatures. *Geological Survey Bulletin* 1452.

Rose, A.W. & G.A. Daub. 1994. Simulated weathering of pyritic shale with added limestone and lime. Proceedings of the International Land Reclamation and Mine Drainage Conference and Third International Conference on the Abatement of Acidic Drainage. *Bureau of Mines Special Publication* SP 06B-94:334-340.

Schafer, R.W., J.J. Cooper & P.G. Vikre, eds. 1988. Bulk mineable precious metal deposits of the Western United States. *Symposium Proceedings, Geological Society of Nevada*. Reno, Nevada.

Singer, P.C. & W. Stumm. 1970. Acid mine drainage: the rate determining step. *Science* 167:1121-1123.

Strömberg, B. & S. Banwart. 1994. Kinetic modelling of geochemical processes at the Aitik mining waste rock site in northern Sweden. *Applied Geochemistry* 9:583-595.

U.S. Department of the Interior. 1994. Proceedings of the International Land Reclamation and Mine Drainage Conference and Third International Conference on the Abatement of Acidic Drainage. *Bureau of Mines Special Publication* SP 06B-94.

Velbel, M.A. 1985. Geochemical mass balances and weathering rates in forested watersheds of the Southern Blue Ridge. *American Journal of Science* 285:904-930.

Wiersma, C.L. & J.D. Rimstidt. 1984. Rates of reaction of pyrite and marcasite with ferric iron at pH 2. *Geochimica Cosmochimica Acta* 48:85-92.

Williamson, M.A. 1992. *Thermodynamic and kinetic studies of sulfur geochemistry*. Ph.D. thesis. Virginia Polytechnic Institute and State University, Blacksburg, Virginia.

Wolery, T.J. 1983. *EQ3NR, a computer program for geochemical aqueous speciation-solubility calculations: user's guide and documentation*. Lawrence Livermore National Laboratory. Livermore, California. UCRL-53414.

Wolery, T.J. & S. A. Daveler. 1989. *EQ6, a computer program for reaction path modeling of aqueous geochemical systems: user's guide and documentation*. Lawrence Livermore National Laboratory. Livermore, California. UCRL draft report.

Tailings & Mine Waste'95 © *1995 Balkema, Rotterdam, ISBN 90 5410 526 7*

Response to the EPA four focused feasibility studies regarding the Summitville mine clean-up by the Technical Assistance Group

Dallas Kent, Jeff Stern, Todd Gilmer, Ken Klco, Mary Mueller, Maya Ter Kuile, Christine Canaly & Wendy Mellott
The Technical Assistance Grant Response

Abstract: This is a community response to the EPA efforts for clean-up at Summitville. The main commentary centers around the absence of water treatment for Acid Mine Drainage at the site and the lack of coordination between clean-up efforts. The reclamation proposal does not address long term conditions that will be prevalent at the site. This paper reviews these concerns and offers practical options.

General Comments on the Approach taken by EPA:

The specific Remedial Action Objectives for all four Focused Feasibility Studies are given at the beginning of each report and are quoted as follows:

*   *"to reduce or eliminate detrimental water quality flow into the Wightman Fork;*
*   *to reduce or eliminate acid mine drainage from man-made sources on the site;*
*   *to reduce or eliminate any human health or adverse environmental effects from mining operations downstream from the site, to include the Alamosa River;*
*   *to reduce or eliminate the need for continued expenditures for water treatment at the site; and*
*   *to encourage early action and acceleration of the Superfund process."*

Despite these stated objectives, there is little attempt to use them to integrate the proposed remedial actions. In contrast, the TAG alternatives outlined in this letter will meet or exceed all the Remedial Action Objectives established for the site by the EPA.

*   The four Focused Feasibility Studies appear to be running parallel courses and little attempt has been made to integrate them with each other.

    *   The studies need to complement each other.

*   We question the use of the Colorado Division of Minerals and Geology for Use Attainability Study of the Alamosa River section between Wightman Fork and Terrace Reservoir to determine validity of State Stream Standards.

    *   Needs Review by *Independent* panel.
    *   We feel there is some bias associated with the Division.

*   All four Focused Feasibility Study use "*best case*" assumptions, which does not account for possible or probable failures.

* This biases choice of remediation methods
* Underestimates long-term capital and O+M costs.

- Downstream impacts are currently being ignored and avoided despite the above stated Remedial Action Objectives.

   * Terrace Reservoir.
   * Agricultural land degradation.
   * Household and municipal wells.

- Need site drainage plan

   * surface/subsurface drainage and storm water management.
   * sedimentation
   * nonpoint source collection/treatment

- Need waste management plan

- The Focused Feasibility Studies should be prepared by third parties without vested interests.

   * Privatize efforts by exploring suggestions from companies in the water treatment and mined land reclamation fields.

- The TAG urges the EPA to avoid the use of "Technical Practicability Waivers" as shortcuts in the remediation because it may set dangerous precedents for future cleanups.

Comments on the Focused Feasibility Study for the Cropsy Waste Pile, Cleveland Cliffs Tailings Pond, Beaver Mud Dump and Mine Pits

- Need Remedial Investigation site characterization plan

   * Characterization and monitoring of ground water flow system is inadequate.
   * Interim ground water modeling results have not been released.
   * Need surface water and sediment monitoring program.

- Need Cleveland Cliffs Tailings Pond incorporated into site storm water management and other surface water plans.

- Design parameters for isolating wastes in Mine Pits are not discussed, and may be inadequate. In other words, what and where are the liners?

   * Underliner
   * Side lining
   * Cap

- Clay liners from on-site materials have Acid Mine Drainage potential

   * How is this potential minimized, eliminated or controlled?

   * Is there enough on-site material that can be used and is it suitable?

   * The same stockpiles of materials are mentioned in the Reclamation Focused Feasibility Study. Is there enough for multiple remedial actions?

36

- Need ground water control plan to prevent inundation of wastes in pit.

  * Figure 6 of the Reynolds Adit Control Program (10/94) shows that the water level has risen above bottom elevation of the pits.

  * In the event of requiring water level control by discharge from the Reynolds Adit, the water removed should be treated (See TAG alternative for water treatment).

- Effectively, a landfill is being created in the Mine Pits.

  * Are State permitting requirements for mine wastes used in the design?

  * Are State permitting requirements for water treatment sludges implemented in the design?

  * Will post-closure monitoring requirements from the State be met?

Comments on the Heap Leach Pad Focused Feasibility Study
In addition to the specific Remedial Action Objectives for the site - stated at the beginning of this document - the *interim* Remedial Action Objectives for the Heap Leach Pad are:
- *to eliminate or minimize Heap Leach Pad impacts to aquatic receptors in Wightman Fork, the Alamosa River and Terrace Reservoir;*
- *to eliminate or minimize the need for continued expenditures for water treatment at the Heap Leach Pad;*
- *to reduce or control Heap Leach Pad drainage so that the Alamosa River will continue to be usable for agriculture in the San Luis Valley;*
- *to reduce or control Heap Leach Pad drainage so that human health will continue to be protected from releases from Heap Leach Pad;*
- *to implement interim remedial action at Heap Leach Pad in an accelerated manner, presumably within two years of signing of the Interim Record of Decision*

EPA's Preferred Alternative (No. 3) is flawed in the following ways:

- There will be both aerobic and anaerobic conditions in the Heap Leach Pad. The Heap Leach Pad will be under water at lower levels but the top will be above water level. This may not be conducive to bacterial activity.

- With the water level in the Heap Leach Pad near the top of Dike 1, spring runoff may exceed the capacity of the water treatment plant to prevent overtopping. Provision should be made for adequate pumping and water treatment capacity to prevent overtopping.

- Cyanide "hotspots" will remain beneath intermediate liners. Untouched cyanide "hotspots" are present in other areas of the Pad because of short circuiting of rinse and caving caused by the rinse. These "hotspots" cannot be reached by a rinse containing bacteria as flows will tend to follow already established pathways.

- The proposed injection wells may become plugged due to biologic activity.

- The French Drain discharge is likely to continue to be a major source of metals and acidity, even after Heap Leach Pad remediation. The drain is connected to the Cropsy Waste Pile, which in turn is fed by seeps and springs. Because the Cropsy Waste Pile is not being removed entirely, there is a continuing source of Acid Mine Drainage to the French Drain.

- Because of significant caving in the Heap Leach Pad caused by continuous rinse, the number and costs of injection wells and their placement is impossible to determine.

- The TAG proposal for water treatment will make the bioreactor unnecessary.

The TAG proposes that Alternative 2 be reassessed and modified:

- Reduced Heap Leach Pad elevation will avoid redox shifts from aerobic to anaerobic conditions and will make treatment more effective.

- The water levels in the Heap Leach Pad will be raised and the remainder of the pad inundated to infiltrate all cyanide sources.

- Removing Heap Leach Pad material below dike level will allow complete inundation of all remaining "hot spots". This material can be moved into the mine pits.

- EPA should use existing application systems and devices, or use exfiltration beds similar to "leach fields" of septic systems to introduce rinse fluids to the Heap Leach Pad.

- The TAG proposes that the effluent from the French Drain be treated until it meets accepted levels or standards. As this is likely to extend past the period contemplated under Alternative 3, the relative value of TAG's water treatment proposal (see following section for its description) is enhanced.

- *Finally, the Heap Leach Pad remediation plan needs to be integrated into the overall site reclamation plan, particularly the storm water management plan, grading plan, and revegetation plan. As has been pointed out, it also needs integration into the water treatment plan.*

Comments on the Reclamation Focused Feasibility Study

- This Focused Feasibility Study is too general and simplistic, even for 30% design level, to expect success.

    * The complex topography, hydrology, and vegetation of the site require the use of differing stabilization and revegetation technologies in separate site areas (microsites).
        - no breakdown of microsite areas in Focused Feasibility Study
        - no reference of specific plant species/seed mixtures for each microsite
        - no reference to soil amendment types and rates for each microsite

    * Reclamation success will be at microsite level

- Preferred Alternative (No. 4) may work over limited site areas but cannot be applied sitewide (refer to Morrison-Knudsen Sitewide Conceptual Remediation Plan).

- Each alternative in the Focused Feasibility Study may be applicable to limited site areas

    * Other alternatives not identified in the Focused Feasibility Study may be needed and applicable.

38

- The Focused Feasibility Study assumes that the capping will work

    * Capping will likely fail in some areas, due to
        - frost action
        - inappropriate cap materials
        - water action/erosion

* Need to plan for failures - The TAG water treatment proposal accommodates potential for failure.
        - drainage and erosion control structures needed
        - sedimentation structures needed

- Need to incorporate test plot results into the Focused Feasibility Study

    * Are the soils in the land application area poisoned by metals?

    * If studies addressing this issue have not been conducted, they need to be done as soon as possible.

- The time frame for remediation is unrealistically optimistic

    * Reclamation will probably require 10 to 20 years in difficult areas

- Costs are likely inaccurate and underestimated

    * Time frame for full revegetation is too short and does not accommodate for potential failures.

- Need a storm water management plan integrated with reclamation plan

    * Grading plan for each microsite
    * Drainage plan for each microsite
    * Grading and drainage plan for whole site
    * Sedimentation control plan for whole site
    * Water treatment plan for storm water runoff from the entire site.

- Need to integrate with other plans.

Comments on Water Treatment Focused Feasibility Study:

Alternative 5 (The preferred EPA alternative) does not accomplish any of the specific Remedial Action Objectives described at the beginning of this document. The TAG proposes a modified version of Alternative 6 which would meet the objectives.

The TAG proposes:

- A water treatment plant to treat ALL acid mine drainage leaving the site. If the EPA's preferred alternative 5 is selected, any flow in excess of 600 gallons per minute (gpm) of acid mine drainage will be discharged untreated directly into the Wightman Fork.

- This plant would be located above the confluence of the Wightman Fork with the Cropsy Creek (see attached recommendations from the Technical Advisors to the TAG).

- The Cleveland Cliffs Tailings Pond will be used for storage of excess runoff in the spring. Upon completion of water treatment, the pond can be converted to a constructed wetland. Because of the limited capacity of the existing treatment plants under the EPA's preferred alternative 5, the pond will not be able to hold excess water and will be constantly disturbed. The TAG alternative allows a stable water body in which development of the constructed wetland will be accelerated.

- The estimated cost based on three quotes from companies that do this kind of work is $8 million. This price includes 4 years of O & M costs (in contrast with the EPA's price tag for this alternative of $32 million).

- Estimated capital costs plus Operations and Maintenance costs to the State Taxpayer would be a maximum of $6 million over the next 5 years. There is a possible cost of $16 million if the EPA's preferred alternative 5 is implemented.

- The TAG alternative can be actively treating all point and non-point sources of acid mine drainage a year earlier than the EPA's preferred alternative 5.

- This plant is totally salvageable and can be moved by the State once the site is reclaimed.

- The duration of water treatment by the proposed plant is flexible and more effective in comparison to that of Alternative 5, given the assumptions of that Alternative.

- In the event of failure of any remedy, the water treatment plant is in place. Failures could occur in:
  * Capping.
  * Adit plugs.
  * Stabilization of Cropsy Waste Pile footprint
  * Unforeseen circumstances.

- The personnel needed for the TAG proposal will be 2 to 3 times less than for EPA's Preferred Alternative 5.

- No on-site contractor will be needed to oversee the plant since operators can be hired by the State or by the construction company.

- Proper placement of this plant will alleviate the need to pump water uphill for water treatment as would be necessary under the EPA's preferred Alternative 5. Pumping would cost $122,000 per year in electrical power alone. The Kilowatt-hour requirement for the pumps *alone* is more than the *entire* TAG proposed plant would require.

- It is suggested that this plant be configured for sulfide precipitation. Hydroxide sludges such as would be produced in the existing water treatment plant have proven to be highly unstable and readily soluble over a wide pH range. *Proper* disposal of these sludges is very expensive (see attached recommendations from the technical advisors to the TAG).

- The TAG alternative would meet or exceed water quality criteria established by the State without having to degrade the Stream Standards.

40

- *The water treatment plant should be the hub for integrating the four focused feasibility studies and for all remedial actions to meet the EPA's stated objectives.*

In contrast to EPA's preferred Alternative 5, the TAG alternative for water treatment will meet or exceed all the Remedial Action Objectives established for the site by the EPA.

References:

The Four Focused Feasibility Studies provided by the EPA.
References too numerous to list at this time due to quantity and time limitation for submittal.

Tailings & Mine Waste'95 © 1995 Balkema, Rotterdam, ISBN 90 5410 526 7

Metal transport between an alluvial aquifer and a natural wetland impacted by acid mine drainage, Tennessee Park, Leadville, Colorado

Suzanne S. Paschke & Wendy J. Harrison
Department of Geology and Geological Engineering, Colorado School of Mines, Golden, Colo., USA

ABSTRACT: An investigation is in progress to characterize metal transport from the St. Kevin Gulch tributary of Tennessee Creek into and through the underlying ground-water system of Tennessee Park near Leadville, Colorado. Low-pH, metal-rich surface water from St. Kevin Gulch recharges the sand and gravel aquifer of the park as the stream emerges from the mountain front and flows across an alluvial fan. In addition, St. Kevin Gulch surface water seasonally recharges a natural wetland in the western portion of Tennessee Park. Oxidizing conditions prevail in the portion of the aquifer underlying the St. Kevin Gulch alluvial fan, and reducing conditions are present where the aquifer is overlain by a natural wetland. Infiltration of acidic surface water from St. Kevin Gulch has lowered the pH of ground water in proximity to the stream. Iron concentrations in the aquifer are also elevated where reducing conditions exist. Additional laboratory analysis and reaction transport computer simulation will provide insight to the hydrologic and chemical processes controlling the fate of acid mine drainage in the Tennessee Park ground-water system.

1 INTRODUCTION

A field investigation is in progress at the St. Kevin Gulch site to assess acid mine drainage impacts to the ground-water system of Tennessee Park. Tennessee Park is an intermontane valley at the northern headwaters of the Arkansas River approximately four miles northwest of Leadville, Colorado. The St. Kevin Gulch mining district is located in the upstream reaches of the St. Kevin Gulch tributary to Tennessee Creek (Singewald, 1955). St. Kevin Gulch receives acidic mine drainage from the mining district and contributes acidic water to Tennessee Creek, the Arkansas River, and ultimately Pueblo Reservoir (Kimball, 1991).

The ground-water system of Tennessee Park consists of a sand and gravel aquifer overlain by a natural wetland in the western portion of the valley. As St. Kevin Gulch emerges from the mountain front, it flows across an alluvial fan prior to reaching the Tennessee Park wetland. This alluvial fan is part of the sand and gravel aquifer and receives significant infiltration of acid mine drainage from

St. Kevin Gulch. During the spring and summer, St. Kevin Gulch flow also reaches the Tennessee Park wetland where it also contributes acidic surface water to the hydrologic system of the valley (Walton-Day, 1991).

Several previous investigations have been performed at the St. Kevin Gulch site by the U.S. Geological Survey Toxic Substance Hydrology Program (McKnight et al., 1988; Kimball et al., 1989; Smith et al., 1989; Bencala et al., 1991; Kimball, 1991; Kimball et al., 1991a; Kimball et al., 1991b; Smith et al., 1991; Walton-Day et al. 1991). These papers focus on the hydrology and geochemistry of the St. Kevin Gulch surface water system as impacted by acid mine drainage (Kimball, in press), including colloidal processes associated with iron transport (Ranville et al., 1988; Ranville et al., 1991).

Previous ground-water investigations at the site focused on the processes controlling active metal accumulation in the Tennessee Park wetland through: 1) construction of flow and metal budgets of surface water and ground water; and 2) characterization of the areal and vertical distribution of metals in the wetland sediments (Walton-Day, 1991). Wetland metal budget results for 1988 and 1989 indicate that iron is the principal metal being removed from surface water by the wetland (Walton-Day, 1991). Iron oxy-hydroxide particulates are deposited in the Tennessee Park wetland from surface water inflow primarily during high spring flow (Walton-Day, 1991). Subsequent to deposition, iron is possibly remobilized through reductive dissolution of ferric oxy-hydroxide or through formation of metal-organic complexes (Walton-Day, 1991). Reduced iron is then transported out of the wetland and into the underlying aquifer during the late summer and fall when downward gradients exist (Walton-Day, 1991). These processes are hypothesized to occur seasonally causing a flushing of iron from the wetland into the underlying aquifer (Walton-Day, 1991). However, limited information is available regarding the ground-water hydrology and geochemistry of the sand and gravel aquifer (Walton-Day and Briggs, 1989; Walton-Day et al., 1990).

Previous computer modeling efforts at the St. Kevin Gulch site have focused on chemical reactions and transport in the surface water system of St. Kevin Gulch (Kimball et al., 1991a; Smith et al., 1991; Broshears et al., 1993). No previous studies have applied a reaction transport modeling approach to the ground-water component of the problem at St. Kevin Gulch.

2   OBJECTIVE

The objective of this study is to quantitatively characterize metal transport from the acid mine drainage source of St. Kevin Gulch to the ground-water system of Tennessee Park. This includes assessing the nature and extent of low-pH, metal-rich water in ground water within the alluvial fan and in the portion of the aquifer underlying the Tennessee Park wetland. The objective is being accomplished by integrating field investigations with reaction transport computer simulations. This paper presents preliminary results of the 1994 field investigation.

3   METHODS

An existing ground-water monitoring well network was expanded by installation
of 18 additional wells during the fall of 1994. Wells were installed and developed
during September and October, 1994, and an initial sampling event was conducted
in October, 1994.

3.1   Well locations and completions

The ground-water monitoring network in Tennessee Parks consists of: 1) thirteen
1-1/4-inch polyvinylchloride (PVC) monitoring wells completed in the sand and
gravel aquifer beneath the Tennessee Park wetland (MW series wells) (Walton-
Day, 1991); 2) nine 1-1/4-inch wetland piezometers for measuring ground-water
levels in the wetland (Walton-Day, 1991); and 3) eighteen two-inch PVC wells
completed in the sand and gravel aquifer as part of this study (TPW series wells)
(Figure 1). Fourteen of the 18 new wells are installed west of the wetland to
characterize ground-water flow and quality within the alluvial fan (wells TPW-1
through TPW-13 and TPW-15). Well TPW-14 is located north of the alluvial fan
to characterize background water quality, and wells TPW-16 through TPW-18 are
located east and south of the wetland to characterize ground-water flow and
quality downgradient of the wetland.

The MW wells and piezometers were drilled with a portable gasoline-powered
hydraulic drilling rig with four-inch diameter solid stem augers (Walton-Day,
1991). The TPW wells were drilled with a truck-mounted CME 55 drilling rig.
TPW boreholes were advanced with 3-1/4 inch or 4-1/4 inch inside diameter (ID)
hollow stem augers, and split spoon samples of the aquifer materials were
collected every five feet. Geologic logs were developed based on split spoon
samples and auger cuttings.

The TPW wells are completed with two-inch outside diameter, Schedule 40,
threaded and flush jointed PVC well casing. The majority of new wells are
completed across the water table; however, deeper wells are paired with the water
table wells at four locations along St. Kevin Gulch (TPW-1/2, TPW3/4, TPW-5/6,
and TPW-7/8) to assess the vertical distribution of metals in the alluvial fan. In
water table wells, a ten foot screened interval is placed across the water table as
encountered during drilling, and a five foot screened interval is submerged below
the water table well screen in the deeper well of well pairs. The screened interval
of all MW and TPW series wells consists of 0.01 inch slotted PVC with PVC
slip-on end caps at the top and bottom of each well. Where not precluded by
borehole caving, 16-40 mesh silica sand was placed in the borehole annulus
around the screened interval in all MW and TPW series wells. A one to two foot
seal of bentonite pellets was placed above the sand pack and saturated prior to
placing the surface seal. The surface seal consists of Portland Type I/II cement
poured to ground surface.

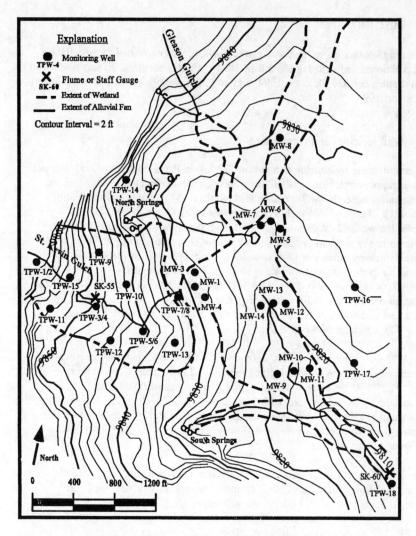

Figure 1. Tennessee Park site map and well locations.

3.2 Well development

All wells were developed subsequent to installation and prior to ground-water sampling to obtain relatively clear ground-water samples. TPW series wells were developed by surging and bailing with stainless steel bailers until relatively clear water was discharged. Repeated bailing was required for wells which bailed dry. For wells which yielded sufficient quantities of water (approximately one gallon per minute), a centrifugal pump was used to pump the wells subsequent to surging and bailing. Relatively clear ground-water samples were obtained from all TPW series wells, indicating that well development was successful.

46

3.3   Ground-water sampling

One round of ground-water samples was collected from all MW and TPW series wells in addition to select surface water stations in October, 1994. Dedicated 3/8" polyethylene tubing was installed in each well, and wells were purged with a peristaltic pump prior to sampling. A minimum of three casing volumes of ground water was pumped from each well immediately prior to sample collection in order obtain representative formation water. Stabilization of pH and conductivity in discharged water was also achieved prior to sample collection.

Ground-water samples were analyzed in the field for pH, temperature, specific conductance, dissolved oxygen, alkalinity, total iron, and ferrous iron (Table 1). Separate sample aliquots were collected for laboratory analyses of dissolved inorganic anions, dissolved metals, total metals (surface water and select ground-water samples only), dissolved organic carbon, and total organic carbon (Table 1). Sample aliquots for dissolved inorganic anions and dissolved metals were field filtered through 0.10 micron (μm) nitrocellulose filters, and the sample aliquot for dissolved organic carbon analysis was field filtered through 0.45 μm glass filters. All sample aliquots were preserved in the field as appropriate for the analysis method (Table 1).

Table 1.   Ground-water and surface water sampling parameters and methods

Parameter	Field preservation	Laboratory method
Field parameters		
pH	unfiltered, unacidified	Electrode
Temperature	unfiltered, unacidified	Thermometer
Specific Conductance	unfiltered, unacidified	Conductivity Cell
Dissolved Oxygen	unfiltered, unacidified	Colorimetric
Alkalinity	unfiltered, unacidified	Gran Titration
Fe^{2+}/Fe^{3+}	unfiltered, unacidified	Colorimetric
Laboratory parameters		
Inorganic ions		
SO_4^{2-}	filtered[1], unacidified	Ion Chromatography
Cl^-	filtered[1], unacidified	Ion Chromatography
HCO_3^-	filtered[1], unacidified	Gran Titration and Ion Chromatography
ICP metals	filtered[1], acidified	Inductively-Coupled Plasma Emission Spectroscopy
Organics		
Dissolved Organic Carbon	filtered[2], unacidified	High Temperature Persulfate Oxidation
Total Organic Carbon	unfiltered, unacidified	High Temperature Persulfate Oxidation

[1] Filtration with 0.1 um nitrocellulose filters.
[2] Filtration with 0.45 um glass filters.

3.4   Laboratory analyses

Laboratory analyses of water samples include inorganic anions, metals, dissolved organic carbon, and total organic carbon (Table 1). Field analyses, dissolved organic carbon, and total organic carbon results are presented in this paper.

4   RESULTS AND DISCUSSION

4.1   Site hydrogeology

The ground-water system of Tennessee Park consists of an unconfined to confined alluvial sand and gravel aquifer. The aquifer materials are primarily composed of very poorly sorted, angular to subrounded, unconsolidated, clayey sands and gravels. Gravel, pebbles, and occasionally cobbles composed of granite, schist, and gneiss rock fragments from the Sawatch Range were encountered in the alluvial fan. Well-sorted fine sand and clay layers are also present in the alluvial sediments. A natural wetland consisting of peat with interbedded fine sand and clay overlies the sand and gravel aquifer in the western portion of the park and is in hydraulic connection with the aquifer (Walton-Day, 1991).

Recharge to the ground-water system occurs as surface water and ground-water flow from the surrounding mountains. Hydrographs of stream flow into and out of the Tennessee Park wetland for 1988 and 1989 indicate that it receives its maximum recharge during the annual spring snowmelt and runoff (Walton-Day, 1991). Maximum surface water recharge occurred in May of both 1988 and 1989, with surface water recharge gradually decreasing throughout the remainder of the year (Walton-Day, 1991).

Surface water flow draining into the wetland is intermittent and ephemeral. St. Kevin Gulch (the primary source of acid mine drainage) is ephemeral and flows only during the spring runoff or in response to storm events during the late summer (Walton-Day, 1991). During September and October 1994, St. Kevin Gulch flow was perennial at a gaging station (SK-40) located approximately 0.6 miles upstream of SK-55. However, the stream loses water to the alluvial fan downstream of SK-40 and has completely infiltrated into the stream bed approximately 150 feet upstream of SK-55. This assertion is consistent with previous observations at the site during the low-flow period (Zellweger and Maura, 1991).

Figure 2 presents the potentiometric surface of the water table in the sand and gravel aquifer for October 1994. Surface water recharge from St. Kevin Gulch is evident where the gulch flows across the alluvial fan, and ground-water flow directions roughly parallel the gulch toward the southeast. However, ground-water flow continues on to the east-southeast and does not bend to the northeast (near well pair TPW-5/6) as does St. Kevin Gulch. This area also appears to be a transition zone from unconfined to confined ground-water flow. A gray plastic

48

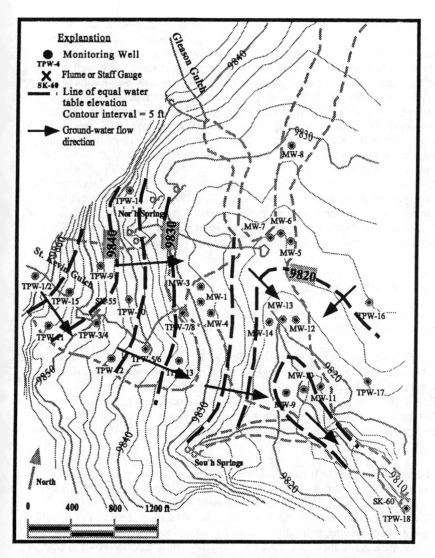

Figure 2.   Elevation of the Tennessee Park water table (in feet above mean
            sea level), October, 1994.

clay was encountered near ground surface in wells TPW-8 and TPW-13. This
clay is approximately six feet thick at TPW-8 and is approximately nine feet thick
at TPW-13. Water levels in TPW-8 and TPW-13 are also indicative of confined
conditions. Ground-water flow directions east of the wetland are toward the
southwest indicating this area is a ground-water divide between the wetland and
Tennessee Creek to the east. Ground-water flow beneath the wetland generally
follows topography toward the southeast.

Similar to surface water flows, ground-water levels in the wetland and the aquifer underlying the wetland are highest during the spring runoff and gradually decline for the remainder of the year (Walton-Day, 1991). In addition, there appears to be a seasonal reversal of vertical hydraulic gradient in the wetland (Walton-Day, 1991). Upward gradients exist during the snowmelt season (May and June for 1988 and 1989) causing upward flow from the aquifer into the wetland, and downward gradients exist during the late summer and fall causing downward flow from the wetland to the aquifer (Walton-Day, 1991). When upward gradients exist (during spring runoff), the wetland receives ground-water discharge from the underlying sand and gravel aquifer, and lateral ground-water flow is received from the surrounding mountains (Walton-Day, 1991). When downward gradients exist (during late summer and fall), the wetland recharges the underlying sand and gravel aquifer (Walton-Day, 1991), and the dominant flow direction in the aquifer is from northwest to southeast toward the wetland confluence with Tennessee Creek. Ground-water discharge also occurs at North and South Springs. Flow at both of these springs is perennial.

4.2   Geochemistry

Limited chemical data are currently available (Table 2); however, these data are consistent with the hydrogeologic conceptual model of the site.

pH is lowest in wells TPW-1 (4.4), TPW-2 (3.9), and TPW-3 (3.9). These wells are in close proximity to St. Kevin Gulch, indicating that low pH impacts to the aquifer are limited in extent. Other notable low pH values occur at TPW-11 (4.9), TPW-6 (4.5), and TPW-13 (4.8). These wells are located along the southeasterly flow path of ground-water emanating from the St. Kevin Gulch drainage. Background pH ranges from 6.0 (North Springs) to 7.0 (TPW-14), and pH values in the wetland range from 5.0 to 6.3.

Dissolved oxygen (DO) was measured at the well head using a colorimetric spectrophotometer with high (greater than 0.880 milligrams per liter [mg/l]) and low range (less than 1.1 mg/l) DO ampule vials. DO levels in wells on the alluvial fan and in surface water levels range from 4.4 mg/l at well TPW-4 to 13.3 mg/l at South Springs, indicative of oxidizing conditions. Split spoon samples of aquifer materials in the alluvial fan are also yellowish to reddish brown in color, again indicative of oxidizing conditions. Reduced conditions are present at well pair TPW-7/8; wells TPW-16, TPW-17, and TPW-18; and at all wells underlying the wetland based on DO measurements of less than 2.4 mg/l in combination with gray sediment color.

Iron concentrations were also measured at the well head using a colorimetric spectrophotometer and appear to correlate with DO measurements. The highest total iron concentrations occur at wells which also exhibit reduced conditions (TPW-7/8 and the wetland wells), except for wells TPW-16 and TPW-17. Iron concentrations are not elevated at TPW-16 and TPW-17 indicating that these wells are not impacted by acid mine drainage from St. Kevin Gulch. This observation

50

Table 2.   Chemical data for ground-water and surface water samples collected from Tenneesee Park, October, 1994.

Station	Temperatur (oC)	pH	Specific conductance (umohs/cm)	Dissolve oxygen (mg/l)	Total iron (mg/l)	Ferrous iron (mg/l)	Total organic carbon (mg/l)	Dissolved organic carbon (mg/l)	Particulate organic carbon (mg/l)	TOC absorance at 400 nm (color units)	DOC absorance at 400 nm (color units)	TOC absorance at 254 nm (color units)	DOC absorance at 254 nm (color units)
TPW-1	8.0	4.39	190	6.4	0.46	0.12	0.62	0.70	-0.09	0.181	0.008	0.288	<0.995
TPW-2	6.0	3.90	295	8.5	0.04	0.00	0.53	0.99	-0.46	0.018	0.008	<0.994	<0.996
TPW-3	8.0	3.86	210	7.8	0.02	0.00	0.54	0.85	-0.31	0.009	0.011	<0.990	<0.990
TPW-4	7.5	5.10	140	4.4	>3.30	0.80	0.62	0.73	-0.11	0.683	0.011	0.862	<0.990
TPW-5	6.5	5.19	250	4.9	0.11	0.00	1.85	2.09	-0.23	0.037	0.013	0.014	0.002
TPW-6	6.5	4.50	179	5.9	0.04	0.00	0.90	0.58	0.32	0.014	0.015	<0.990	<0.990
TPW-7	7.5	6.29	175	1.1	>16.50	>16.50	1.04	2.83	-1.79	0.091	0.113	0.350	0.159
TPW-8	5.5	6.80	150	2.4	2.25	2.06	3.60	1.67	1.93	0.022	0.016	0.042	0.024
TPW-9	6.0	5.79	98	5.7	0.36	0.10	2.17	4.36	-2.19	0.065	0.028	0.19	0.025
TPW-10	7.0	5.92	141	7.4	2.15	0.36	1.29	4.79	-3.51	0.023	0.017	0.005	0.006
TPW-11	5.5	4.88	110	7.0	0.00	0.00	1.04	1.50	-0.47	0.028	0.013	<0.990	<0.993
TPW-12	5.9	5.25	139	6.4	0.02	0.00	0.89	1.40	-0.51	0.017	0.018	<0.990	<0.996
TPW-13	6.5	4.78	115	5.1	0.12	0.10	1.02	2.72	-1.70	0.014	0.016	<0.990	<0.994
TPW-14	6.0	7.02	94	6.1	1.03	0.39	4.36	1.50	2.86	0.264	0.092	0.127	0.146
TPW-15	7.0	5.19	100	6.1	0.07	0.00	0.61	1.46	-0.85	0.016	0.017	<0.990	<0.990
TPW-16	7.0	5.10	220	1.8	0.01	0.00	2.83	1.15	1.68	0.015	0.014	0.005	0.005
TPW-17	6.5	5.00	150	1.5	0.24	0.00	1.36	0.87	0.50	0.015	0.012	<0.992	<0.990
TPW-18	5.8	6.04	77	0.426	0.19	0.18	0.87	3.20	-2.34	0.015	0.025	nd	0.054
MW-1	4.9	5.52	121	0.88	3.19	2.92	1.10	2.98	-1.87	0.032	0.022	0.116	0.075
MW-3	5.9	6.32	123	0.1	>3.30	>3.30	0.96	1.64	-0.67	0.142	0.186	0.256	0.334
MW-4	3.9	5.81	115	2.3	0.22	0.00	0.71	0.80	-0.09	0.141	0.014	0.161	<0.995
MW-9	7.5	5.00	150	0.728	1.36	0.80	1.11	1.32	-0.21	0.017	0.013	0.004	0.002
MW-10	6.5	5.52	210	0.201	>3.30	>3.30	3.91	4.99	-1.09	0.063	0.096	0.224	0.260
MW-11	6.5	5.50	120	0.213	>3.30	>3.30	3.43	4.23	-0.80	0.027	0.042	0.111	0.18
MW-12	6.0	5.26	108	1.2	0.03	0.01	1.80	1.04	0.76	0.017	0.019	<0.995	0.017
MW-13	6.2	6.20	141	0.263	>3.30	>3.30	12.20	12.55	-0.35	0.137	0.159	1.249	1.494
MW-14	6.5	6.31	135	0.116	>3.30	>3.30	2.77	4.75	-1.97	0.098	0.059	0.163	0.437
SK-40	8.0	3.62	235	9.7	1.62	0.12	0.92	1.02	-0.10	0.031	0.030	0.027	0.019
SK-CULV	5.5	3.90	298	9.9	1.52	0.16	0.83	1.50	-0.67	0.022	0.020	0.008	0.001
N. Springs	4.9	6.01	92	9.4	1.03	0.02	3.61	1.62	1.98	0.017	0.014	0.018	<0.996
MW10SW	6.8	5.96	162	9.3	0.22	0.05	3.88	4.88	-0.99	0.028	0.022	0.079	0.078
S. Springs	8.5	6.60	40	13.3	0.14	0.05	0.92	1.90	-0.98	0.016	0.020	<0.995	0
SK-60	2.5	6.35	60	9.1	0.26	0.05	2.43	1.94	0.49	0.019	0.025	0.009	0.022

is consistent with water table elevations which imply that the subtle topographic ridge east of the wetland is a ground-water flow divide. Walton-Day (1991) hypothesizes that dissolved iron from the wetland is transported to the underlying aquifer, the current data set appears to support this conclusion.

Dissolved organic carbon (DOC) values range from 0.73 mg/l at TPW-4 to 12.55 mg/l at MW-13. In general, the lowest DOC concentrations (less than one mg/l) occur at wells which appear most impacted by low pH water (TPW-1/2, TPW-3/4, and TPW-6). Higher DOC concentrations (greater than four mg/l) occur in wells completed beneath the wetland with the maximum DOC concentration occurring at MW-13 near the center of the wetland. Total organic carbon (TOC) concentrations were, in general, slightly higher than DOC concentrations. This anomalous result may indicate biodegradation of the unfiltered TOC sample aliquots.

5   SUMMARY AND CONCLUSIONS

Field investigations and computer simulations are in progress to characterize the nature and extent of acid mine drainage impacts to the Tennessee Park ground-water system. Initial field work is complete, and laboratory analysis of ground-water and surface water samples is in progress. The ground-water system of Tennessee Park consists of an unconfined to confined sand and gravel aquifer. Oxidizing conditions prevail in the portion of the aquifer underlying the St. Kevin Gulch alluvial fan, and reducing conditions are present where the aquifer is overlain by a natural wetland. Infiltration of acid mine drainage impacted surface water from St. Kevin Gulch has lowered the pH of ground water in proximity to the stream. Where reducing conditions exist, iron concentrations in the aquifer are also elevated. Additional laboratory analyses and reaction transport computer simulation of ground-water flow and metal transport will provide a tool for understanding the hydrologic and chemical processes controlling the fate of acid mine drainage emanating from St. Kevin Gulch.

6   ACKNOWLEDGEMENTS

The activities on which this report is based were financed in part by the Department of the Interior, U.S. Geological Survey, through the Colorado Water Resources Research Institute. Field and laboratory support is also provided by the U.S. Geological Survey Toxic Substance Hydrology Program. The contents of this publication do not necessarily reflect the views and policies of the Department of the Interior, nor does mention of trade names or commercial products constitute their endorsement by the United States Government. Funding for reaction transport computer model development and simulation is provided, in part, by the Colorado Advanced Software Institute, Platte River Associates, Inc, and PTI Environmental Services.

REFERENCES

Bencala, K.E., B.A. Kimball, and D.M. McKnight, 1991, Use of variation in solute concentration to identify interactions of the substream zone with instream transport: U.S. Geological Survey Water-Resources Investigations Report 91-4034, pp. 377-379.

Broshears, R.E., K.E. Bencala, B.A. Kimball, and D.M. McKnight, 1993, Tracer-dilution experiments and solute-transport simulations for a mountain stream, Saint Kevin Gulch, Colorado: U.S. Geological Survey, Water-Resources Investigations Report 92-4081, 18 p.

Kimball, B.A., 1991, Physical, chemical, and biological processes in waters affected by acid mine drainage: from headwater streams to downstream reservoirs: U.S. Geological Survey Water-Resources Investigations Report 91-4034, pp. 365-370.

Kimball, B.A., in press, Effects of seasonal variations on metal concentrations in acid mine drainage, St. Kevin Gulch, Colorado in: Plumlee, G.S. and Logsdon, M.H., eds., Reviews in Economic Geology, v. 6, Society of Economic Geologists.

Kimball, B.A., K.E. Bencala, D.M. McKnight, 1989, Research on metals in acid mine drainage in the Leadville, Colorado area: U.S. Geological Survey, Water-Resources Investigations Report 88-4420, pp. 65-70.

Kimball, B.A., R.E. Broshears, K.E. Bencala, and D.M. McKnight, 1991a, Comparison of rates of hydrologic and chemical processes in a stream affected by acid mine drainage: U.S. Geological Survey Water Resources Investigations Report 91-4034, pp.407-412.

Kimball, B.A., D.M. McKnight, G.A. Wetherbee, and R.A. Harnish, 1991b, Mechanisms of iron photoreduction in metal-rich, acidic streams: Chemical Geology, in press.

McKnight, D.M., B.A. Kimball, and K.E. Bencala, 1988, Iron photoreduction and oxidation in an acidic mountain stream: Science, v. 240, pp. 637-640.

Ranville, J.F., K.S. Smith, D.L. Macalady, and T.F. Rees, 1988, Colloidal properties of flocculated bed material in a stream contaminated by acid mine drainage, St. Kevin Gulch, Colorado: U.S. Geological Survey Water-Resources Investigations Report 88-4220, pp. 111-118.

Ranville, J.F., K.S. Smith, D.M. McKnight, D.L. Macalady, and T.F. Rees, 1991, Effect of organic matter coprecipitation and sorption with hydrous iron oxides on electrophoretic mobility of particles in acid mine drainage: U.S. Geological Survey Water-Resources Investigations Report 91-4034.

Singewald, Q.D., 1955, Sugar Loaf and St. Kevin Mining Districts Lake County, Colorado: U.S. Geological Survey Bulletin 1027-E, pp. 251-299.

Smith, K.S., D.L. Macalady, and P.H. Briggs, 1989, Partitioning of metals between water and flocculated bed material in a stream contaminated by acid mine drainage near Leadville, Colorado: U.S. Geological Survey Water Resources Investigations Report 88-4420, pp. 101-110.

Smith, K.S., J.F. Ranville, D.L. Macalady, 1991, Predictive modeling of copper, cadmium and zinc partitioning between streamwater and bed sediment from a stream receiving acid-mine drainage, St. Kevin Gulch, Colorado: U.S. Geological Survey Water-Resources Investigations Report 91-4034, pp. 380-386.

Walton-Day, K., 1991, Hydrology and geochemistry of a natural wetland affected by acid mine drainage St. Kevin Gulch, Lake County, Colorado: Ph.D. Thesis T-4033, Colorado School of Mines, 299 p.

Walton-Day, K. and P.H. Briggs, 1989, Preliminary assessment of the effects of acid mine drainage on ground water beneath a wetland near Leadville, Colorado: U.S. Geological Survey Water-Resources Investigations Report 88-4420, pp. 119-124.

Walton-Day, K., P.H. Briggs, and S.B. Romberger, 1991, Use of mass flow calculations to identify processes controlling water quality in a subalpine wetland receiving acid mine drainage, St. Kevin Gulch, Colorado: U.S. Geological Survey Water-Resources Investigations Report 91-4034, pp. 371-376.

Walton-Day, K., D.L. Macalady, M.H. Brooks, and V.T. Tate, 1990, Field methods for measurement of ground water redox chemical parameters: Ground Water Monitoring Review, v. 10, n. 4, pp. 81-90.

Zellweger, G.W. and W.S. Maura, 1991, Calculation of conservative tracer and flume discharge measurement on a small mountain stream: U.S. Geological Survey Water-Resource Investigations Report 91-3034.

Tailings & Mine Waste'95 © 1995 Balkema, Rotterdam, ISBN 90 5410 526 7

Superfund listing of mining sites

C. A. Patton
Woodward-Clyde, Sacramento, Calif., USA

K. M. McGaffey
Bogle and Gates, Seattle, Wash., USA

J. L. Ehrenzeller, R. E. Moran & W. S. Eaton
Woodward-Clyde, Denver, Colo., USA

ABSTRACT: The Hazard Ranking System (HRS) was developed by the U.S. Environmental Protection Agency (EPA) as a screening tool to evaluate the relative risk to human health and the environment posed by sites with uncontrolled releases of hazardous substances. In theory, the application of the HRS model should result in uniform technical judgements by EPA regarding a site's potential hazard and, an approximation of risk; however, the HRS can be biased toward listing metal mining sites on the National Priorities List (NPL). The model often ignores or does not require accurate characterization of risk related-factors typically evaluated in formal risk assessments. These limitations of the HRS model are especially problematic at mining sites where such factors significantly influence the calculation of potential risk posed by a site. Therefore, it is important for Potentially Responsible Parties (PRPs) to be actively involved in EPA's evaluation from data collection activities through the public comment period once the draft HRS scoring document is available for public comment and technical review.

1 INTRODUCTION

This paper provides a brief summary of the Superfund process, examines special considerations for mining sites (particularly the apparent bias in the process toward listing mining sites on the National Priorities List [NPL]), and recommends several actions mining companies can take to participate in the Superfund process and respond effectively in the event their facility is proposed for inclusion on the NPL.

The Comprehensive Environmental Response, Compensation and Liability Act of 1980 (CERCLA/Superfund) (PL 96-510) requires that criteria be established based on relative risk or danger, in order to establish guidelines used to evaluate facilities where hazardous substances may have been released. These criteria must account for the population at risk; the hazardous potential of the substances at a facility; the potential for contamination of drinking water supplies, direct human contact, and destruction of sensitive ecosystems; and other risk-related factors.

The Hazard Ranking System (HRS) incorporates the criteria required by CERCLA and is used as a screening tool by EPA to identify sites requiring further investigation under provisions of the Superfund process. The HRS is the principle means by which EPA determines whether it should place sites on the NPL.

The HRS was originally adopted in the July 16, 1982 Federal Register (47 FR 31180) as Appendix A of the National Oil and Hazardous Substances Pollution Contingency Plan (NCP). When Congress passed the Superfund Amendments and Reauthorization Act of 1986 (SARA), amending CERCLA, Congress directed EPA to amend the HRS scoring system. As a result, EPA proposed revisions in the December 23, 1988 Federal Register (53 FR 51962) and finalized the revisions in the December 14, 1990 Federal Register (55 FR 51532).

The revised HRS model provided a number of substantive changes to the original scoring procedure. In addition to revising each factor used in the original system, some new factors were added. The three basic components of source, pathway, and receptor remained central to the ranking system.

Factors evaluated in the HRS were intended to approximate both the probability of harm from a facility and the magnitude of harm that could result. The HRS ranks facilities in terms of the potential threat they pose by describing:
- Characteristics and quantity of the harmful substances.
- Possible migration pathways (routes) and the likelihood of hazardous substance release.
- Potentially affected targets.

The HRS assigns a score from 0 to 100 to a site based upon the potential for harm to human health or the environment from migration of a hazardous substance or from exposure at the source. The system may assess one or more of the following four pathways:
- Groundwater Migration
- Surface Water Migration
- Soil Exposure
- Air Migration

The score for each migration pathway is obtained by considering a set of factor categories that characterize the potential of the facility to cause harm. These factor categories include Likelihood of Release, Waste Characteristics, and Targets. Each factor within a factor category is assigned a numerical value according to HRS guidelines. Scores are added within each factor category, then the total scores for each category are multiplied together to develop a score for groundwater, surface water, air, and soil exposure. The scores for each pathway are combined into a total site score. A total score of 28.5 or greater is sufficient to qualify a site as a "proposed site" to be considered for inclusion on the NPL.

2 NPL LISTING PROCEDURES

Following the identification of a site by EPA, or state agencies, or citizens who petition EPA to consider a site, sites are entered into the Comprehensive Environmental Response, Compensation, and Liability Information System (CERCLIS). This is EPA's computerized inventory of sites where releases of hazardous substances are possible. The pre-remedial activities of the NPL process consist of a series of investigations which evaluate, in increasing detail, whether a CERCLIS site is likely to qualify for the NPL. A site moves from one stage to the next based on the likelihood that it will qualify for the NPL. In the first stage of this process, Preliminary Assessments (PAs) are conducted. PAs focuses primarily on obtaining available information or "desktop" data for key HRS scoring factors which enable the evaluator to derive a projected HRS score. Based on the anticipated HRS score, the evaluator recommends no further action or a medium or high priority Site Inspection (SI). If no further action is recommended the site is no longer evaluated in the pre-remedial process.

SIs are conducted to build upon information obtained during the PA phase. SIs collect additional data (including environmental samples to further characterize the site, and specific data relating to key HRS scoring factors) and verify data previously collected during the PA. Sampling is conducted at the SI stage to identify the types of contaminants present, to determine if a release of contaminants has occurred, and to assess the likelihood that actual human or environmental exposure to contaminants has occurred. Occasionally, more than one SI is conducted to gather additional data.

Once the necessary data have been collected through the PA and SI stages of the pre-remedial process, a complete HRS scoring package is prepared by EPA or its contractors. It is estimated that only 3 to 5 percent of the sites included on CERCLIS become the subject of formal HRS scoring. The HRS score is subject to quality control review at the regional EPA level. Quality assurance audits for accurate and consistent application of the HRS, as well as appropriate eligibility criteria, are then conducted at EPA headquarters. Sites passing these reviews and receiving a HRS score of 28.5 or greater can be proposed for inclusion on the NPL. A list of proposed sites to be added to the NPL is published in the Federal Register. EPA then requests public comment on the proposed sites and accepts comments for 60 days following the publication in the Federal Register. Extensions of the 60-day comment period can occur but are rare. Following the 60-day comment period, EPA considers all relevant comments and adds to the NPL (Final Rule) all proposed sites that continue to meet the NPL criteria for listing. Finally, EPA prepares a Support Document which provides an explanation of the detailed rationale applied by EPA in calculating the HRS score. The NPL process can often take several years to complete.

3 CRITIQUE OF THE HRS MODEL

In theory, the application of the HRS model should result in (1) uniform technical judgements by EPA regarding a site's potential hazard and (2) an approximation of actual risk; however, the HRS can be biased toward listing some types of sites, mining sites in particular, on the NPL. A close examination of the HRS model, comments produced during the revisions to the HRS, and comments provided in challenges to EPA's nomination of sites to the NPL, indicates that EPA is willing to accept a number of tradeoffs and some loss of precision in order to use the HRS as a screening tool rather than an assessment of actual risk. EPA's continued use of the HRS model with its inherent limitations is the primary cause of the bias toward the listing of mining sites on the NPL.

As described by EPA in the Agency's Response to Comments on the revisions to the HRS:

"The objectives of the final HRS require considerable simplification of the detailed and complex calculations performed in formal risk assessments. This simplification results in some loss of precision. The HRS is not, and was never intended to be, a risk assessment. The agency has employed numerous simplifying assumptions within the final HRS that are required to permit relative ranking of sites based on the information collected during preliminary assessments and site inspections. The Agency believes that the final HRS reflects formal risk assessment principles to the maximum extent feasible and that the loss of precision resulting from simplification is acceptable" (EPA, no date, p. 1A-5).

Many scientists, consultants, attorneys, and mining industry representatives disagree with EPA's position and have aggressively challenged this basic premise. Throughout the history of the HRS, and especially during the revisions to the HRS required by SARA, the HRS model has been heavily criticized for the apparent bias created by these simplifying assumptions toward assigning high scores to mining sites. While it is commonly believed that the inability of the model to account for high volume/low toxicity wastes at mining sites is primarily responsible for the apparent bias, comments received by EPA on the revised model and responses to proposed NPL listings of mining and smelter sites, indicate that this bias is the result of a combination of factors including:

- Failure of the model to evaluate certain technical factors that are commonly used in the evaluation of actual risk
- Structure (or mathematics) of the HRS algorithm
- Procedural requirements for the application of the HRS

Each of these factors is discussed below.

3.1 Technical Factors

Although it is understood that the HRS is not intended to be a formal risk assessment commensurate with EPA guidelines for such evaluations, it is clear that EPA's simplification of the process for the purpose of screening can lead to overestimation of the actual risk at some sites. The model fails to adequately consider background concentrations, contaminant characteristics, and groundwater flow direction and these technical problems often lead to the overestimation of actual risk at mining sites.

3.1.1 Background Concentrations

Accurate establishment of background concentrations is critical to the HRS process; background concentrations established below the detection limits for the analytical method used can result in the calculation of an Observed Release based on the mere detection of an analyte in a sample. This is particularly important at mining sites where naturally occurring levels of metals are expected. By definition mining sites are located in highly mineralized areas where "elevated" concentrations of metals and other constituents existed prior to the onset of mining activities. Significant fluctuations in these background concentrations over time may be expected due to seasonal changes or the location of a sample with respect to the orebody and local and regional geologic structures. Given the expected natural variability, the evaluation of large data sets (if available) or the use of modeling techniques are often necessary to produce a scientifically defensible estimation of background concentrations. However, EPA is often unwilling to review large data sets or accept the results of modeling in order to evaluate background concentrations (Krause, 1994). As a result,

EPA's data sets and literature searches are often incomplete resulting in inadequate and scientifically indefensible background concentrations.

For example, in the recent proposed listing of a mining site for which substantial amounts of data were available regarding background samples for the groundwater pathway, EPA failed to evaluate the natural variability demonstrated by the large data set (EPA, 1994). Instead, EPA established background concentrations based on selected samples which showed no detection of various metals even though other sampling events for the same wells showed levels above the detection limits of the analysis. This approach allowed EPA to establish an Observed Release for the site based upon the mere detection of metals in selected groundwater samples. Without the proper technical evaluation of background concentrations or procedures to check the reasonableness of EPA's assessment (see Section 3.3) it is entirely possible that an HRS score greater than 28.5 could be generated for a site based on concentrations of metals that are routinely found in mineralized areas prior to mining activities.

3.1.2 Contaminant Characteristics

Toxicity, mobility, persistence and bioaccumulation values used in the HRS process are established by the Superfund Chemical Data Matrix (SCDM). SCDM is an EPA-generated source which provides factor values and benchmark values for contaminants frequently found at sites. These values generally assume the most toxic, most persistent, and most bioaccumulating species of a compound are present at a site, unless other species of the same compound are also listed (EPA, 1993). This assumption can significantly overestimate the risk posed by a mining site since these parameters cannot be accurately defined for metallic ore minerals without consideration of the speciation of the compound.

Appropriate toxicity values for metals (specifically lead, arsenic, and boron) have been the subject of significant debate. In the National Gypsum case the court rejected EPA's unsupported use of toxicity values for this particular site (see Section 6.1). Although, the outcome of this case is favorable to the mining industry, there is no indication that issues of arsenic and lead will be decided any time soon, and it is clear that all cases will be considered on a purely case-by-case basis.

Aside from the controversy over toxicity values, the issue of mobility of contaminants in the groundwater pathway is perhaps one of the most hotly contested issues with respect to mining sites. The Scientific Advisory Board (SAB) has noted, "[n]atural geochemical processes are often required to mobilize hazardous substances in mining waste, such as acid generation by the biochemical oxidation of metal sulfides" (EPA no date, p. 12I-51). Without regard to speciation and other site-specific conditions, the HRS assumptions regarding the mobility of metals in the groundwater pathway can greatly overestimate potential for contaminants to be released from a site and, therefore, the risk posed by a site.

The bioavailability values assigned by SCDM are also a point of contention. Metals derived from mining and milling activities, and metals within the soil matrix are generally less bioavailable to human or environmental receptors, and, therefore, pose less risk than indicated by the values provided in SCDM. Since the HRS does not consider the reduced bioavailability of metals from many mining-related sources, the risk at mining sites can be overestimated by the HRS.

SCDM also establishes benchmark values for various compounds. These values are used to weight target populations that have been affected by a release (as defined by the HRS) from a site. Much weight is given to those populations that have been exposed to contaminant levels greater than the benchmark value provided in SCDM. However, for some metals that may be found at mining sites (including lead), SCDM fails to incorporate current scientific knowledge pertaining to toxicology, exposure, and risk in assigning the benchmark values. As a result, benchmark values established by the HRS can be inconsistent with similar health-based screening levels which EPA regions use to identify sites that pose no unacceptable risk and require no further action. Therefore, EPA's use of HRS benchmark values (which may be inappropriately low) in the scoring process can result in the listing of a site where other EPA assessments would establish the site poses no unacceptable risk. For example, at several mining sites EPA has established clean-up levels for metals during the remedial investigation and feasibility stages of the CERCLA process that are greater than the benchmark values used in the HRS.

3.1.3 Groundwater Flow Direction

Groundwater flow direction is not considered in the HRS model. Although it is highly unlikely that releases to groundwater are equally likely to flow in all directions, the HRS considers all targets within four miles of the sources identified at a site. The HRS does give more weight to targets that have actually been impacted by a release from the site, however, it does not prevent the calculation of an HRS score greater than 28.5 based solely upon groundwater targets that are located upgradient from the source of contaminants. Since the upgradient targets are not likely to be affected by releases or potential releases, the HRS overestimates the risk posed by a site.

3.2 Structural Factors

Primary components of a formal risk assessment evaluation and an HRS evaluation are similar, and include the source of contaminant, route of contaminant migration, and target (or receptor). However, the mathematical and procedural rules of each algorithm are substantially different. The mathematical implications of the HRS are discussed in this section. The implication of the procedural rules are discussed in Section 3.3.

One of the most significant differences in the mathematical process of the algorithms is that a formal risk assessment requires and establishes a direct link between sources, routes of migration, and targets, while the HRS does not. For example, the HRS allows the score for a site to be evaluated based on (1) the waste characteristics of a large well-contained Source A, (2) the likelihood of a release from a small poorly-contained Source B, and (3) the targets associated with Source C. This approach clearly overestimates the actual risk posed by a site and is especially problematic at mining sites where a number of sources are typically located over a large geographical area. In responding to comments on the revisions to the HRS, EPA actually agreed that this approach could result in the overestimation of risk; however, the agency stated that given the uncertainties in site inspection data, it was

60

uncertain whether a more ideal approach would result in more accurate site scores, therefore the simpler approach was adopted in the HRS (EPA, no date, p. 1C-1).

The mathematics of the algorithm used to calculate HRS scores have been questioned both in challenges to proposed listing and comments on the revisions to the HRS. Rather than discuss the merits of these additional challenges and comments, the remainder of this section describes the "reality" of the mathematics for mining sites.

The mathematics of the model often result in qualifying HRS scores for mining sites which have high volume/low toxicity waste. However, this apparent bias is related to both the manner in which Waste Characteristic values (including toxicity) are established in the HRS process (see Section 3.1) and the fact that mining sites also often receive relatively high values for the Likelihood of Release category. When combined, the values for these two categories require relatively little contribution from the Target category in order to produce a score which would qualify a site for the NPL. For example, a Target category value of only 85 is required to qualify a site for the NPL based on a single pathway which receives maximum values for both the Likelihood of Release category and the Waste Characteristics category. An even smaller target category value (8.5) is required for the Surface Water Pathway, where the maximum value for the Waste Characteristics category is 1000 (compared to 100 for the other migration pathways). Thus, although mining sites may be located in isolated areas, only a few targets may be necessary in order to qualify the site for inclusion on the NPL.

3.3 Procedural Factors

Despite EPA policies designed to promote consistency and quality throughout the HRS process on a national scale, the technical accuracy, level of documentation, clarity of presentation, and general quality of HRS packages varies widely among EPA regions and with each proposed rule. Although EPA asserts that peer review procedures and the establishment of regional decision teams in the Superfund Acceleration Cleanup Model (SACM) have improved the quality and consistency among HRS packages, a comparison of recently proposed HRS packages suggests the HRS lacks the necessary checks and balances to ensure quality and to evaluate the reasonableness of the scores assigned.

The HRS is commonly criticized as a "black box" into which both an experienced HRS scorer or an inexperienced HRS scorer can place the required information, perform the HRS calculations and produce an HRS score without regard for the reasonableness of the assigned value. When calculating an HRS score, many instances arise in which an experienced risk assessor would use their knowledge of chemical and site-specific characteristics to make a technical judgement. However, these opportunities are circumvented in the HRS by the use of "look-up" tables and step-by-step directions. In such situations, the inexperienced HRS scorer (or experienced HRS scorer lacking adequate time and resources) may opt to rely on the less accurate default values and the step-by-step directions provided by the model, while the more experienced scorer with time and resources may apply discretion and consider or compensate for the models limitations. Amax Mining Co. compared the possibility of variable results in the HRS to the variable results observed in an exercise in tax form preparation sponsored by Money Magazine in 1989. Amax pointed out that the most important observation was that the flexibility of scoring suggests that the basic approach developed by EPA is likely to be open to a wide range of interpretations and application and that assessing accuracy would be problematic and difficult at best (EPA, no date, p. 1A-12). For example, EPA's scoring of the Blackbird Mine in May 1993 specifically excluded the open pit as a source of contamination while EPA's scoring for another open pit mine in January 1994 included the open pit without justification.

Although EPA has published guidance and policies regarding the importance of quality and consistency in preparing HRS scores, one significant document cited by the HRS Guidance manual has yet to be promulgated by the Agency. This document is entitled the "Data Useability Guidance for Superfund Assessment." When published, the document should describe the appropriate collection, interpretation, and useability of chemical analysis data required to support the scoring of sites under the HRS (EPA 1992, p. 8). Without this document, the HRS has no substantive requirement that data collected by PRPs meet the same stringent requirements established for data collected by EPA. Yet, in the response to comments on the revised HRS, EPA stated that the Agency believes that "proper quality assurance applied to HRS field activities and data review will eliminate any bias against former mining sites" (EPA, no date, p. 5X-7). Because EPA has failed to establish the parameters of "proper" quality assurance, data collected by PRPs for purposes entirely different than the HRS (such as compliance monitoring) can be used by EPA to document an HRS score without regard for the importance of the quality and reliability of the data.

4 HISTORY OF MINING SITES ON THE NPL

Approximately 39 western U.S. mining sites have been proposed for inclusion on the NPL since 1981. This figure is based upon the independent review described below. Although only four mining sites have been proposed for listing since the promulgation of the revised HRS (December 1990), it is impossible to attribute the slight decrease in the average number of mining sites proposed each year to any particular factor. Contributing factors to EPA's selection of NPL candidate sites may include the effects of budget reductions and policy considerations (such as Resource Conservation and Recovery Act [RCRA] site eligibility and Bevill waste exclusion), or the possible interest of EPA in the listing of mining sites.

According to EPA's NPL Characterization Project of 1991, the mining industry represented three percent of the sites included on the NPL in 1991 (EPA, 1991, p. 32). Based on the total number of sites on the NPL in 1991, this equates to a total of 37 mining sites on the NPL at that time. More recent characterization studies are not available through EPA. However, an independent review of various data sources including the CERCLIS database and numerous Records of Decision, indicates that between 1981 and May 1994 approximately 39 mining sites in the western U.S. had been proposed for the NPL. Thirty-one of the 39 sites were formally listed on the NPL as of May 1994. Although the review does not include the eastern U.S., it does suggest that the majority of mining sites proposed for the NPL become NPL sites despite the active participation of PRPs in challenging proposed listings.

5 CHALLENGING NPL LISTINGS

5.1 Judicial Challenges to HRS Listing Decisions

Judicial challenges to EPA decisions to add sites to the NPL are filed in the United States Court of Appeals for the District of Columbia (D.C.) (see 42 U.S.C. 9613[a]). Although the NPL currently contains more than 1200 sites, the D.C. Circuit Court has decided less than fifteen NPL listing cases (see D.C. Circuit Court cases spanning from 1985 to 1993). A review of these decisions reveals an extreme reluctance on the part of the court to interfere with EPA's delegated authority to design and use the HRS model as a screening tool for evaluating the relative risk posed by a site. In recent years, however, the court has shown a greater willingness to scrutinize EPA's technical application of the HRS model to particular sites.

Since EPA began adding sites to the NPL, parties have challenged the underlying assumptions and structure of the original and the revised HRS models; however, the D.C. Circuit Court has rejected these challenges. For example, the court has rejected challenges to EPA's refusal to take remedial actions into account (D.C. Circuit Court 1985, 1991a), EPA's method of calculating waste quantity (D.C. Circuit Court 1991a), and EPA's method of calculating distances between sources and targets (D.C. Circuit Court 1992f).

Challenges to a variety of specific technical issues regarding EPA's application of the HRS model to a particular site are also common. For years, the D.C. Circuit Court had rejected these challenges as well. Beginning in 1991, however, the D.C. Circuit Court began to look more closely at the science underlying EPA's decisions to add sites to the NPL. In Anne *Arundel County v. U.S. EPA* (D.C. Circuit Court 1992e) and *Kent County v. U.S. EPA* (D.C. Circuit Court 1992a), the court rejected EPA's use of only unfiltered samples when applying the HRS model. In *National Gypsum Company v. U.S. EPA* (D.C. Circuit Court 1992b), the court rejected EPA's unexplained use of toxicity values based upon a highly toxic form of boron compounds when only the less toxic form of boron oxide had been disposed of at the site. In *Tex Tin Corporation v. U.S. EPA* (D.C. Circuit Court 1991c), the court rejected EPA's unsupported assumptions about the source of arsenic-laden dust to score the site.

5.2 Preparing Challenges to NPL Listings

Many challenges by PRPs to proposed NPL listings begin with the notification of the listing in the Federal Register. However, the preparation of a comprehensive, well-reasoned challenge to a potential NPL listing is a labor-intensive process that requires more preparation time than the 60-day comment period provided by EPA. It is recommended that PRPs become actively involved in all aspects of EPA's evaluation of their facility. This involvement should begin as soon as EPA shows an interest in a site. Ideally, PRP involvement should begin during EPA's data collection activities for the PA, and should continue with the additional data collection and field activities for the SI, and through the 60-day public comment period once the HRS scoring package is made available for public comment and technical evaluation.

While EPA is conducting a PA or SI, a PRP or representative familiar with the HRS process may obtain information integral to the preparation of a challenge to a proposed listing. This information may be related to the procedures used by EPA in the collection of data or related to EPA's rationale and strategy for the identification of targets and sources, the selection of background and release sample locations, the attribution of contaminants to a source, and the documentation of releases. A thorough understanding of this information and different scoring strategies can then be used to develop and evaluate the strengths and weaknesses of possible scoring scenarios which may be considered by EPA. Evaluation of these scenarios prior to the proposed listing of a site greatly facilitates and expedites the preparation of a comprehensive, well-reasoned and technically supported challenge. This proactive approach is especially important at mining sites where the complexity of the relationships between various site-specific factors are more significant than at most other types of sites.

In many instances, successful challenges to the proposed listings of mining sites require the coordinated efforts of a skilled multidisciplinary team. The input from both legal and technical teams is essential to provide comprehensive comments regarding the regulatory history of the site; the legislative and judicial background of CERCLA; the factual and technical accuracy of the HRS package; and the compliance of EPA's actions with CERCLA, HRS guidance and sound scientific principles. It is recommended that these multidisciplinary teams be established early in the CERCLA process and expanded as necessary relative to EPA involvement and the likelihood that a site might qualify for the NPL.

6 SUMMARY

EPA's simplification of standard risk assessment procedures for the purpose of screening and establishing the relative risk among sites using the HRS model results in the overestimation of the actual risk posed by some sites (especially mining sites). Limitations regarding technical, structural, and procedural aspects of the model contribute to the bias toward assigning high scores to mining sites. These limitations are especially problematic at mining sites where the accurate assessment of site-specific factors is essential for the accurate evaluation of risk. PRPs can take several steps to ensure that risks are accurately assessed during the HRS process and to facilitate and expedite a well-reasoned, comprehensive response to a proposed NPL listing (should this become necessary). These steps include early involvement in EPA's evaluation of a site, evaluation of scoring scenarios prior to the proposed listing, and the assembly of an experienced multidisciplinary team. Regulators can also take actions to reduce the model's bias against mining sites, primarily through (1) a commitment to agency-wide procedures (such as peer review and decision teams) which ensure the uniformity and technical accuracy of HRS packages, and (2) the consideration of past EPA actions at other mining sites (such as the completion of risk assessments and the establishment of clean-up goals) when setting priorities for listing sites on the NPL.

REFERENCES

D.C. Circuit Court. 1985. *Eagle-Picher Industry, Inc. v. U.S. EPA.* Federal Register. 759 F.2d & 921.

D.C. Circuit Court. 1987. *Eagle-Picher Industry, Inc. v. U.S. EPA.* Federal Register. 822 F.2d 132.

D.C. Circuit Court. 1988a. *Northside Sanitary Landfill, Inc. v. Thomas.* Federal Register. 849 F.2d 1516.

D.C. Circuit Court. 1988b. *City of Stoughton v. U.S. EPA.* Federal Register. 858 F.2d 747.

D.C. Circuit Court. 1990. *Washington State Department of Transportation v. U.S. EPA.* Federal Register. 917 F.2d 1309.

D.C. Circuit Court. 1991a. *Linemaster Switch v. U.S. EPA.* Federal Register. 938 F.2d 1299, 1307 & 1321.

D.C. Circuit Court. 1991b. *B & B Corporation v. U.S. EPA.* Federal Register. 938 F.2d 1299.

D.C. Circuit Court. 1991c. *Text Tin Corporation v. U.S. EPA.* Federal Register. 925 F.2d 1321.

D.C. Circuit Court. 1992a. *Kent County v. U.S. EPA.* Federal Register. 963 F.2d 391.

D.C. Circuit Court. 1992b. *National Gypsum Company v. U.S. EPA.* Federal Register. 968 F.2d 40.

D.C. Circuit Court. 1992c. *Apache Powder Company v. U.S. EPA.* Federal Register. 968 F.2d 66.

D.C. Circuit Court. 1992d. *Bradley Mining Company v. U.S. EPA.* Federal Register. 972 F.2d 1356.

D.C. Circuit Court. 1992e. *Anne Arundel County v. U.S. EPA.* Federal Register. 963 F.2d 412.

D.C. Circuit Court. 1992f. *B & B Tritech, Inc. v. U.S. EPA.* Federal Register. 957 F.2d 822 & 884.

Krause, A. J., and R.M. Eddy. 1994. "The Failure of CERCLA in Addressing Mine Waste" in Tailings and Mine Waste '94.

U.S. EPA. No date. Responses to Comments on the Revisions to the Hazard Ranking System. Washington, D.C.

U.S. EPA. 1991. NPL Characterization Project: National Results. Office of Emergency and Remedial Response. Washington, D.C. November.

U.S. EPA. 1992. Hazard Ranking System Guidance Manual. U.S. Department of Commerce, National Technical Information Service. Springfield, VA. November.

U.S. EPA. 1993. Superfund Chemical Data Matrix (SCDM). March.

U.S. EPA. 1994. Documentation Record for Kennecott (South Zone). January.

Tailings & Mine Waste'95 © 1995 Balkema, Rotterdam, ISBN 90 5410 526 7

Simulation of pit closure alternatives for an open pit mine

Hannah F. Pavlik, Fred G. Baker & Xiaoniu Guo
Baker Consultants, Inc., Golden, Colo., USA

James S. Voorhees
Santa Fe Pacific Gold Corporation, Albuquerque, N.Mex., USA

ABSTRACT: A numerical ground water flow model and particle tracking module were developed for the design of a mine dewatering system and prediction of post-closure hydrogeologic conditions in a series of mine pits proposed in a Tertiary basalt flow sequence in the western United States. Two general pit closure alternatives were evaluated. The first approach allows ground water levels in the vicinity of the pits to recover to a new equilibrium leading to the formation of lakes in the two deepest pits. The second approach requires backfilling of the pit bottoms with permeable materials so as to prevent ground water from rising above the upper surface of the backfill, thereby preventing lake formation.

Simulation of the pit lake alternative showed that ground water levels in and around the pits would be lowered considerably below the original water table elevation. Evaporative losses from the lake surfaces would act as hydraulic sinks causing ground water to flow toward and into the pit lakes. Although the inflows could lead to changes in the chemistry of the lake water, the hydraulic gradient would be directed primarily into the lakes, minimizing regional ground water impacts. Backfilling the pits would cause significantly less lowering of the local water table. However, ground water recharge entering the backfilled materials within the pits could potentially migrate into the surrounding country rock.

1 INTRODUCTION

A hydrogeologic investigation was conducted to evaluate the environmental impacts of a proposed open pit mining operation in the western United States. The mine site is located in the Basin and Range physiographic province in an area dominated by north-northwest trending mountain ranges comprised of Tertiary basalt flows gently dipping toward the southeast. The ranges are separated by relatively flat-lying drainage basins consisting of rift valleys filled with alluvial and colluvial sediments that were deposited by permanent and ephemeral streams.

The climate at the proposed mine site is temperate and arid. The mean annual precipitation is 8.26 inches per year based on 43 years of record (NOAA, 1992). Precipitation is found to increase with elevation as is typical in the Basin and Range region. Evaporation during the summer months is high, reaching a maximum of approximately 24 inches per year.

Baseline hydrologic and water quality conditions were established to support mine planning decisions and to complete the engineering design of a pit dewatering system and a well field required to supply supplemental water for mine and mill operations. In addition, two general pit closure alternatives were evaluated to anticipate the potential impacts of mine closure on planned mining operations and on surface and ground water resources after mining has ceased.

The first closure option allows ground water levels in the vicinity of the pits to recover to a new equilibrium leading to the formation of shallow lakes in the two deepest pits. The second closure option requires backfilling of the pits with permeable rock and overburden materials so as to prevent ground water from rising above the upper surface of the backfill, thereby preventing lake formation. Evaluation of the environmental impacts of the two pit closure scenarios is a critical element of mine planning and permitting activities. Mining practices that minimize the drawdown of the local ground water table need to be evaluated against practices that minimize potentially adverse impacts to surface and ground water quality.

2 CONCEPTUAL MODEL OF THE HYDROGEOLOGIC SYSTEM

Three major structural geologic events shaped the complex hydrogeologic conditions that currently exist in the vicinity of the proposed mine pits: (1) the Roberts Mountains Overthrust which occurred in the mid Paleozoic, (2) the development of the Northern Nevada Rift which was responsible for the formation of a series of large, northwest-striking grabens associated with extensive faulting, dike emplacement, volcanism, and mineralization during the Miocene, and (3) Basin and Range normal faulting in the late Miocene (Stewart et al., 1977). As a result of these events, the local stratigraphic column consists of Quaternary alluvial/colluvial deposits overlying a considerable thickness (3,000 to 4,000 feet) of Tertiary volcanics that rest unconformably on a basement of Paleozoic cherts, quartzites and greenstones. The Tertiary volcanic sequence consists of capping basalts, basaltic andesites and dacites, and a variety of pyroclastic lithologic units.

The conceptual hydrogeologic model for the site indicates that ground water flows outward from ground water recharge areas located near the crest of the mountains just a short distance west of the proposed mine pits. Ground water flows down-dip through the volcanic bedrock aquifer system and discharges into unconfined and confined aquifer units located downgradient in the surrounding valleys and basins. The permeability of the basalt bedrock is generally low (approximately 0.014 feet per day) and dominated by layers with limited but effective fracture porosity. The storativity of the bedrock is also low (on the order of 10^{-5}), making the bedrock aquifer highly sensitive to ground water recharge. In contrast, the alluvial valley-fill units present in the adjacent basins are moderately to highly permeable (mean hydraulic conductivity of 10 to 26 feet per day) with high specific yield (0.1).

Hydrogeologic conditions are complicated by the existence of abundant faults and several natural freshwater springs that are located in the vicinity of the area proposed for excavation. Several high-angle normal faults and fault zones filled with dikes or fault gouge are present. During the Miocene these features acted as conduits for the injection of hydrothermal fluids into the country rock and formed the mineralized zones targeted for mining. Today these faults and fault zones are filled with highly-altered, very low-permeability clays and zones of silica replacement (hydraulic conductivity of

10^{-5} to 10^{-8} feet per day). Fracture porosity in bedrock on either side of the faults is discontinuous due to the presence of these low permeability zones which act as barriers to ground water flow and cause hydraulic gradients to steepen around the mine pits. This effect can be observed on the baseline water table map shown in Figure 1. In some cases, as much as 300 feet of displacement exists between the water table observed on the upgradient and downgradient sides of major fault zones.

Numerous freshwater springs occur on the upland slopes surrounding the proposed mine pits. The majority of the springs represent the surface expression of a single, local ground water potentiometric surface. Spring elevations occur at or just below this surface. Spring locations tend to correlate with the presence of subcropped layers of relatively highly permeable fractured or vesicular lava flows. This suggests that fracture networks or vesicular zones present within the bedrock aquifer are highly interconnected and readily conduct ground water to the spring discharge points.

3 GROUND WATER FLOW MODELING

A detailed three-dimensional ground water flow model was developed for assessment of hydrologic baseline conditions, design of the optimum mine pit dewatering system, and evaluation of post-closure hydrogeologic conditions near the mine pits. The model was based on the USGS Modular Three-Dimensional Ground Water Flow Model (MODFLOW) (McDonald and Harbaugh, 1988) and was used to represent a 18.7 square mile area in 10 layers of finite-difference blocks each measuring 250 feet by 250 feet. The blocks within each model layer were assumed to have uniform thickness and were assigned aquifer hydraulic properties that were equivalent to the hydrogeologic unit encountered at any given block location. In addition, the particle-tracking module PATH3D (Zheng, 1989) was used to evaluate the direction and approximate rate of ground water flow in the vicinity of the mine pits under both post-closure scenarios.

The detailed flow model was used to simulate the elevation of the ground water table after steady-state conditions have been reestablished after mine closure. The planned maximum extent of excavation of the mine pits was assumed for both the backfilled and non-backfilled cases. The long-term change in the water table surface due to mine closure was estimated by comparing the water table elevations that were initially present under baseline (pre-mining) conditions with those predicted to be present after complete recovery of the water table has taken place. The rate of recovery of the water table to form the pit lakes was estimated from the rate of inflow to the pits at the end of active mining operations and at the post-closure steady-state condition. The baseline configuration of the ground water table that existed before mining is shown in Figure 1.

3.1 Simulation of the backfilled mine pits

Model simulations were performed to evaluate the impact of the backfill closure option on the local hydrologic system, assuming that the mine pits are backfilled with relatively permeable rock or overburden materials. Ground water recharge through the backfill was assumed to occur at the same rate as that expected of infiltration through native soils. This assumption appears to be reasonable because native soils present in the vicinity of the mine pits are generally granular with a relatively high infiltration

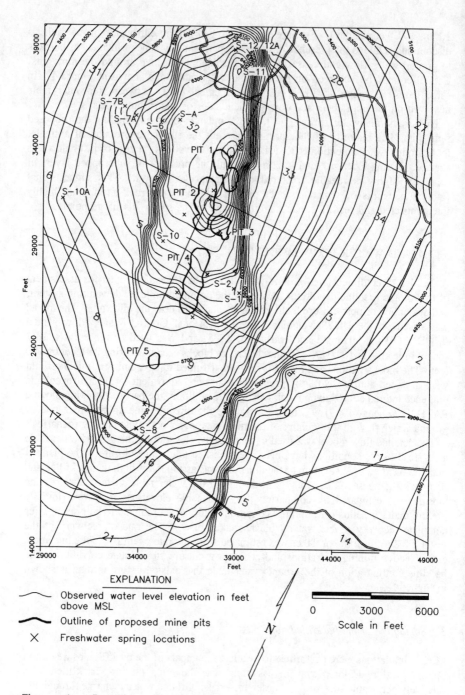

EXPLANATION

⌒ Observed water level elevation in feet above MSL

━ Outline of proposed mine pits

✕ Freshwater spring locations

0   3000   6000

Scale in Feet

N

Figure 1.   Baseline ground water table.

capacity, properties that might be expected of backfill composed of rock and overburden.

In contrast, the backfill was assumed to be more porous and permeable than the natural basalt lying in place in the bedrock aquifer. The porosity and hydraulic conductivity of the backfill were estimated to be 0.10 and two feet per day, respectively. Finally, the backfill was assumed to rise at least 20 feet above the final ground water elevation within each pit.

The post-closure ground water table that is predicted to develop in mine pits that will be closed by backfilling is illustrated in Figure 2A. In all of the proposed mine pits, the post-closure water table is expected to lie beneath the surface of the backfill. Comparison of this water table surface map with the initial baseline water table (Figure 1) indicates that the general form of the post-closure water table will remain relatively unchanged as a result of backfilling except in some local areas around Pits 2, 3, and 4. At the ground water divide located near the crest of the mountains (west of the pits), the post-closure water table is predicted to be approximately 15 feet lower than that present under baseline conditions. In general, an effective decline of 10 to 40 feet will be observed in the mine pit area under the backfilled-pit scenario. This represents a slight flattening of the overall water table surface.

The relatively small decline that will be observed in the water table after closure of the mine pits by backfilling is not expected to adversely impact the majority of the natural freshwater springs that will remain after mining operations have ceased. Reduced flows may be evident at spring S-10A which lies in an area where the water table is expected to be lowered by approximately 15 feet. However, the change in the elevation of the water table near most of the other springs (S-1, S-2, S-6, S-7A, S-7B, S-8, S-11, S-12, S-12A and S-A) is likely to be on the order of zero to ten feet, which is not expected to significantly impact their occurrence or discharge.

3.2 *Simulation of the open mine pits*

The detailed ground water flow model was also used to simulate the post-closure, steady-state ground water conditions that would be established in the event that the pits excavated during mining would not be backfilled upon mine closure. In these simulations, each open pit was represented by a material of relatively high porosity and permeability (1.0 and two feet per day, respectively) to account for the significantly greater water storage capacity of the pits relative to the surrounding bedrock.

Under this closure scenario, it is expected that permanent lakes will form in Pits 2 and 4 when the ground water table has reached equilibrium. A 20-foot deep lake is expected to develop in the southernmost depression in Pit 2 in approximately 10 to 20 years. In addition, a 110-foot deep lake is expected to form in the southern portion of Pit 4. This pit is predicted to refill in approximately 90 to 140 years. The depths predicted for both pit lakes represent long-term annual averages. In the short term, the lake levels are expected to vary by plus or minus 20 feet due to single storm events or seasonal fluctuations in precipitation or evaporation.

The ground water modeling results indicate that post-closure ground water elevations around Pits 2 and 4 will be lower when the mine pits are left open and pit lakes are allowed to form (Figure 2B) than if the pits are backfilled (Figure 2A). The additional decline predicted in the water table is due to the relatively high evaporative losses that are expected from the surface of the pit lakes. In the vicinity of Pits 1 and 3, the water

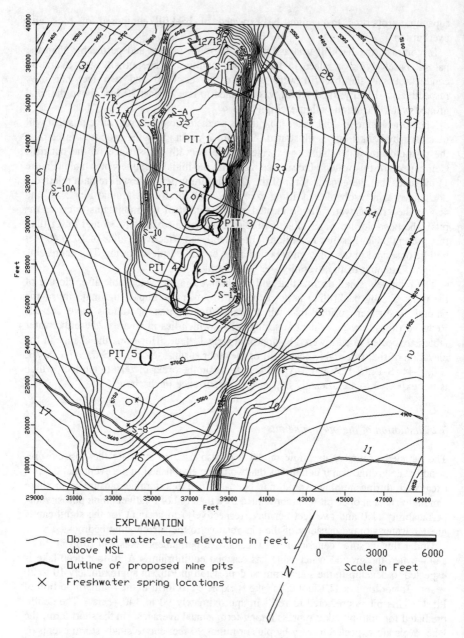

Figure 2A.   Post-closure ground water table in backfilled mine pits.

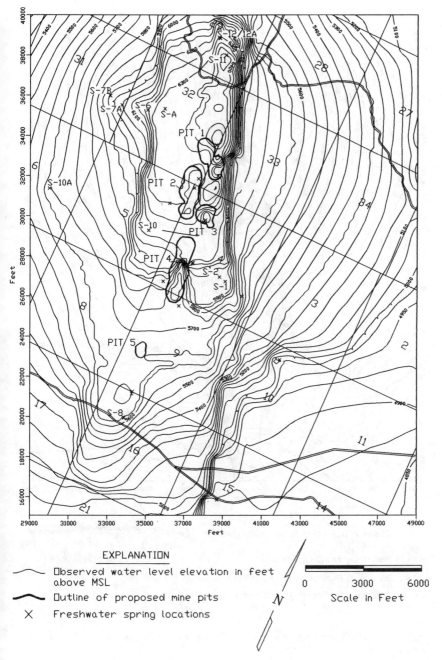

Figure 2B. Post-closure ground water table in open mine pits.

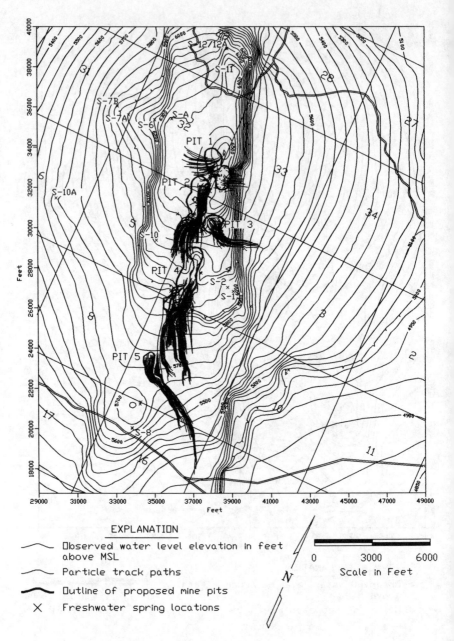

EXPLANATION

⌒ Observed water level elevation in feet above MSL

⌒ Particle track paths

━ Outline of proposed mine pits

✕ Freshwater spring locations

0   3000   6000
Scale in Feet

N

Figure 3A. Particle release from backfilled pits 100 years after closure.

74

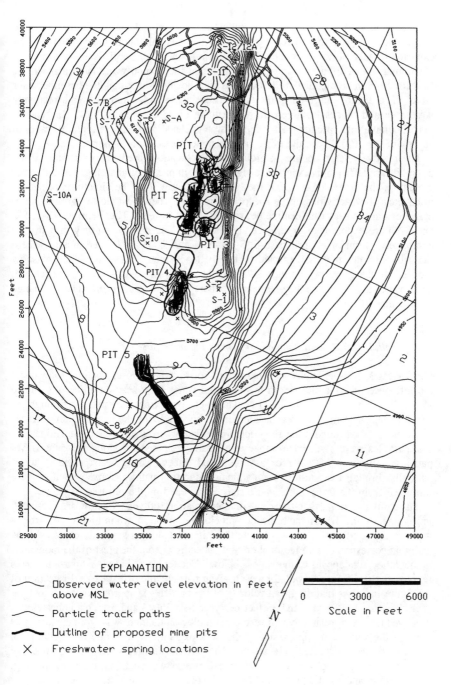

EXPLANATION

⌒ Observed water level elevation in feet above MSL

⌒ Particle track paths

━ Outline of proposed mine pits

X Freshwater spring locations

0   3000   6000
Scale in Feet

Figure 3B. Particle release from open pits 100 years after closure.

75

table is expected to fall to elevations below the base of each pit. This indicates that no permanent lakes will develop in these mine pits, although some water can be expected to accumulate seasonally due to storm events and runoff from snow melt.

Comparison of Figures 1 and 2B indicates that the ground water table that will be established in mine pits that are left open is significantly lower than the baseline water table. The decline observed in the post-closure water table extends over a relatively large area and ranges from zero to a maximum of 280 feet near Pit 4. In general, model simulations predict that ground water elevations will fall an average of approximately 100 feet in the mine pit area under this closure option.

The relatively large decline in ground water elevations predicted to develop if the mine pits are left open is expected to significantly impact natural springs located in the vicinity of the excavated pits. Of the 11 springs that will not be directly impacted by mine excavation activities, three are located in areas where post-closure ground water elevations are expected to decline between 70 and 110 feet. These springs (S-1, S-2 and S-A) are expected to experience significantly reduced flows or they may dry up completely. Four other springs (S-6, S-7A, S-8, and S-11) are located in areas where the ground water table is expected to fall 10 to 30 feet. These are likely to show reduced discharge rates. Springs S-7B, S-10A, S-12, and S-12A are not expected to be impacted at all because they lie in areas where the water table is predicted to change by less than 10 feet.

3.3 *Potential migration of contaminants from the mine pits*

Particle-tracking simulations were performed for both post-closure options to evaluate the influence of mine closure on local ground water flow near the mine pits and to assess the potential for migration of contaminants away from the mine site through the ground water system. This was accomplished by simulating the introduction of tracer particles into the ground water flow field immediately surrounding the closed mine pits. The particles were assumed to move at the same velocity and along the same travel paths as molecules of water or as unretarded tracers would be expected to move naturally through the aquifer. Contaminant migration pathways and velocities were estimated from the travel paths and distances travelled by the released particles.

The results from particle-tracking simulations after 100 years of flow are presented in Figures 3A and 3B. In the backfilled-pit closure option shown in Figure 3A, ground water is expected to flow outward from the mine pits along a generally eastward flow path that corresponds with regional ground water flow from the crest of the mountains to the valley-fill deposits of the valleys below. The length of individual particle tracks represents the distance travelled by each given particle after 100 years. Inspection of the figure indicates that the mean velocity of the particles is approximately 15 feet per year near Pits 1, 2, and 3 and 33 feet per year near Pits 4 and 5. This indicates that dissolved tracers or relatively non-reactive solutes would move away from the mine pits through the aquifer at approximately the same velocities. The rate of migration of reactive solutes is expected to be lower. The particle-tracking simulations do not take into account other physical and chemical processes, such as dilution, mixing, precipitation, and sorption that would also take place along the ground water flow path.

The results from the particle-tracking simulation performed for the open-pit post-closure option after 100 years of ground water flow are shown in Figure 3B. In this case, ground water is predicted to flow primarily inward toward the lakes that have formed in Pits 2 and 4. The reversal in the ground water gradient around these two pits is caused by the local inflow of ground water to the pit lakes from the surrounding aquifer to replenish water lost from the lake surface by evaporation. In effect, the large evaporative losses from the surface of each lake create a local ground water sink. Outside of the immediate drawdown zone associated with each pit lake, ground water flow is largely unaffected by lake evaporation. The ground water flow field that develops around the Pits 1 and 3 (which are not expected to refill with permanent lakes) is predicted to be similar to the flow field that will develop if these pits are backfilled, but less outward migration of ground water is expected to occur. This is because the elevation of the recovered water table is close to the elevations of the pit bottoms allowing some evaporation to occur and thereby limiting outward ground water flow. Ground water flow from Pit 5 will not be affected by evaporation and will largely reflect pre-mining conditions.

4 SUMMARY AND CONCLUSIONS

The results from this work suggest that there is a tradeoff to be achieved between the use of hydraulic controls or geochemical controls to minimize the impacts of mining activities on the water resources of this site. Evaluation of the tradeoff is necessary for sound decision-making. If backfilling is exercised as a pit closure option, relatively little drawdown of the local ground water table can be expected. No permanent pit lakes will form after mine closure and the decline in the water table will cause reduced flows at one existing spring. However, ground water will flow outward from the mine pits into the regional ground water system, establishing a potential pathway for the migration of contaminants.

In contrast, if the mine pits are left open after mine closure, permanent lakes will develop in Pits 2 and 4. Ground water elevations in the vicinity of the pit lakes will be significantly lower than under baseline conditions, causing flows from seven adjacent springs to be adversely impacted. The local reversal in ground water gradients observed adjacent to the lakes will direct ground water flow into the lakes and effectively control the release of potential contaminants from Pits 2 and 4 into the regional ground water system. Less outward migration of ground water is expected to occur from Pits 1 and 3 under this pit closure scenario but potential contaminant migration from Pit 5 may occur.

5 REFERENCES

McDonald, M.G. and A.W. Harbaugh. 1988. A modular three-dimensional finite-difference ground water flow model. Techniques of Water Resources Investigations of the U.S. Geological Survey, Book 6, Ch. A-1.
National Oceanic and Atmospheric Administration (NOAA). 1992. Climatological data summary. ISNN 0364-5312.
Zheng, C. 1989. PATH3D. A ground-water path and travel-time simulator. S.S. Papadopulos & Associates, Inc.

Inventorying and characterization of mine wastes through remote sensing: The Cripple Creek mining district

Douglas C. Peters
US Bureau of Mines, Denver, Colo., USA

Phoebe L. Hauff
Spectral International, Inc., Lafayette, Colo., USA

K. Eric Livo
US Geological Survey, Denver, Colo., USA

ABSTRACT: Non-coal abandoned mine lands have been identified as a significant problem in the U.S., especially because estimates of their number and extent vary widely and often have limited field surveys to back them up. Because this is a high-visibility issue within the U. S. Department of Interior, the U.S. Bureau of Mines, in cooperation with the U.S. Geological Survey and other agencies, is researching techniques for inventorying and characterizing mine wastes to allow more accurate and complete estimates of the waste problem. Remote sensing techniques are particularly attractive because they offer a relatively rapid method for making reproducible estimates of the number and extent of mined lands over large areas. Conversely, standard field updating of existing estimates, or even initial data collection, can be a long, labor-intensive and costly process.

To evaluate remote sensing technology applied to mine wastes, the Cripple Creek mining district was chosen as an initial study area for this research project. The area is attractive due to availability of a number of remote sensing data types and well known geology and history of mining operations. This presentation will cover results of the research to date, such as 1) identifiability of wastes on satellite and aircraft images and 2) mineralogical characteristics of the wastes important for remote sensing, based on field and laboratory analyses. Planned additional work on the Cripple Creek District and at another field site also will be discussed.

SELECTED REFERENCES

Brown, R.L. 1991. Cripple Creek--Then and now. Denver: Sundance Publ. Ltd.

Earth Satellite Corporation 1971. Remote sensing for mined area reclamation: Applications inventory (Report #PB 204 197). Washington: National Technical Information Service.

Gott, G.B., J.H. McCarthy, Jr., G.H. VanSickle & J.B. McHugh 1969. Distribution of gold and other metals in the Cripple Creek District, Colorado. U.S. Geological Survey, Professional Paper 625-A.

Grimstad, W.N. & R.L. Drake 1983. The last gold rush. Victor, Colorado: Pollux Press.

Goetz, A.F.H. 1987. AVIRIS: The new future in geologic remote sensing. In Proc. Pecora XI Symp.: p.88. Falls Church, Virginia: American Society for Photogrammetry and Remote Sensing.

Kenny, J.F. & J.R. McCauley 1982. Remote sensing investigations in the coal fields of southeastern Kansas. In C.J. Johannsen & J.L. Sanders (eds.), Remote sensing for resource management: p.338-346. Ankeny, Iowa: Soil Conservation Society of America.

Lee, K. (ed.) 1989. Remote sensing in exploration geology (28th International Geological Congress Field Trip Guidebook T-182). Washington: American Geophysical Union.

Livo, K.E. 1994. Use of remote sensing to characterize hydrothermal alteration of the Cripple Creek area, Colorado. M.Sc. Thesis, Colorado School of Mines, Goilden, Colorado.

Munts, S.R., P.L. Hauff, A. Seelos & B. McDonald 1993. Reflectance spectroscopy of selected base-metal bearing tailings with implications for remote sensing. In Proc. of 9th Them. Conf. on Geol. Remote Sensing: p.567-578. Ann Arbor: Environmental Research Institute of Michigan.

Peplies, R.W., N.S. Fischman & C.F. Tanner 1982. Detection of abandoned mine lands: A case study of the Tug Fork basin. In C.J. Johannsen & J.L. Sanders (eds.), Remote sensing for resource management: p.362-376. Ankeny, Iowa: Soil Conservation Society of America.

Rowan, L.C. & E.H. Lathram 1980. Mineral exploration. In B.S. Siegal & A.R. Gillespie (eds.), Remote sensing in geology: p.553-605. New York: John Wiley & Sons.

Taranik, D.L. & F.A. Kruse 1989. Iron mineral reflectance in geophysical and environmental research imageing spectrometer (GERIS) data. In Proc. of 7th Them. Conf. Remote Sensing for Expl. Geol.: p.445-458. Ann Arbor: Environmental Research Institute of Michigan.

Thompson, T.B. (ed.) 1986. Cripple Creek mining district. Denver Region Exploration Geologists Society, Fall Field Guidebook.

Tailings & Mine Waste'95 © 1995 Balkema, Rotterdam, ISBN 90 5410 526 7

Recovery of water quality in a mine pit lake after removal of aqua-cultural waste loading: Model predictions versus observed changes

Henry (Hal) M. Runke & Charles (Chico) J. Hathaway
Barr Engineering Company, Minneapolis, Minn., USA

ABSTRACT: Barr Engineering Company prepared a comprehensive lake restoration plan for Minnesota Aquafarms, Inc., a firm involved in net-pen aquaculture operations in an abandoned iron mine pit in northern Minnesota that had filled with groundwater to become a lake. Several years of aquaculture operations had degraded the water quality of this lake from near pristine to extremely eutrophic (i.e., highly fertile), with frequent algal blooms and severe depression of near-bottom dissolved oxygen concentrations. Phosphorus was found to be the element controlling algal bloom frequency and severity, while decaying algae was depleting the bottom lake waters of their oxygen. Therefore, Barr's restoration plan recommended the use of aluminum sulfate to chemically precipitate the phosphorus out of the lake water. The plan also recommended subsequent burial of the precipitated phosphorus and accumulations of fish feces and uneaten fish food with a layer of inorganic soil excavated hydraulically from the mine pit walls.

In addition to prescribing restorative techniques for the improvement of lake water quality, Barr designed measures to manage lake water quality in an adjacent mine pit lake. Specifically, we designed waste collectors for installation beneath the net pens to collect any uneaten fish food and fecal matter that falls through the pens and would otherwise be deposited on the lake bottom. Wastes collected in this manner are now pumped to a waste storage lagoon from which they are spray-irrigated onto agricultural fields. This system, in combination with the addition of ferric chloride on a continuous drip treatment basis, now maintains good water quality in the previously polluted lake.

1 INTRODUCTION

Minnesota Aquafarms, Inc. (MAI) operated a trout and salmon net pen aquaculture facility in the Twin City-South mine pit lake (Figure 1) since June 1988. Over this time period, total phosphorus concentrations in the lake rose from under 10 µg P/L to over 115 µg P/L. Based upon stipulation agreement between MAI and the Minnesota Pollution Control Agency (MPCA), aquaculture operations were phased out of the Twin City-South mine pit lake, and consolidated with the existing operations in the Sherman mine pit lake by July 1, 1993. The stipulation agreement required MAI to implement a demonstration project to restore water quality in the Twin City-South mine pit lake over the next several years. It specified that

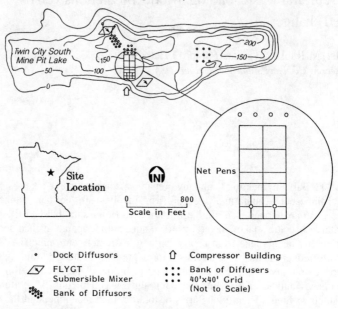

Figure 1

restoration activities must be designed to restore the water quality to a level that is approximated by the following individual target measures by 1996:

- Total phosphorus (growing season mean) - 10 µg/L or less [in the lake's photic zone].
- Dissolved oxygen - no presence of biogenic meromixis or anoxia [i.e., dissolved oxygen concentration < 0.1 mg/L] in the lower half of the hypolimnion, excluding the lowest three meters above the lake bottom.
- Chlorophyll a (growing season monthly mean) - 3 µg/L or less.
- Secchi disc transparency (growing season mean) - 3 meters or greater.

The stipulation agreement limited 1993 lake restoration activities to the use of aeration and artificial circulation methods optimized to oxidize the organic material in the lake sediments. This paper discusses both the results of the 1993 aeration activities, and the restorative methods recommended for use in the Twin City-South mine pit lake during 1994 and 1995 to meet stipulated 1996 water quality levels.

2 TWIN CITY-SOUTH MINE PIT LAKE

2.1 Basin Formation

Mine pit lakes, including the Twin City-South mine pit lake, are vastly different from natural lakes in many respects. Whereas most natural lakes in Minnesota were formed by glacial activities 10-12,000 years ago, the Twin City-South mine pit lake was excavated by man to extract iron ore from an ore body embedded in the bedrock

82

of the area. Earliest iron mining efforts consisted of underground tunneling activities that included vertical shafts and horizontal drifts that followed the underground veins of ore. Subsequently, when strip mining became feasible, the glacial till (overburden) overlying the ore-containing bedrock was stripped away and piled around the mine pit perimeter. The iron ore was then excavated out of the bedrock in open-pit fashion. To accomplish this, the mine pit had to be constantly dewatered to remove inflowing groundwater. This dewatering lowered the surrounding water table level to as much as 400 feet below the surrounding surficial groundwater levels.

2.2   Water Sources

Since mining activities ceased at the Twin City-South mine pit, inflowing groundwater has been filling the lake basin at a rate that currently causes the lake surface to rise about 5-feet per year, on average. Direct precipitation and runoff from a very small watershed also add minor amounts of water to the lake each year. Equilibration of the lake surface with the surficial groundwater table is not expected to occur for 20 years or more.

2.3   Physical Structure

The Twin City-South mine pit lake is extremely deep (and getting deeper in response to groundwater inflows), has a relatively small surface area and short wind fetch, and is sheltered by the surrounding topography. It is unlikely to mix completely each spring and autumn, and is expected to become a naturally meromictic lake, irrespective of any aquacultural activities conducted there. Meromictic lakes do circulate at times, but incompletely. In contrast to holomictic lakes, the entire water volume of a meromictic lakes does not participate in the mixing. A dense stratum of bottom water remains characteristically stagnant and anaerobic. There are several causes of meromixis in lakes, but the MPCA feels that meromixis in the Twin City-south mine pit lake, if/when it occurs, would be of biogenic origin, caused by aquaculture. In biogenically meromictic lakes there is an accumulation of substances from bacterial decay of organic matter that diffuses out of the sediment and into the overlying lake waters.

An example of a naturally meromictic mine pit lake similar to the Twin City-South Lake is the Hartley mine pit lake, about 3/4 mile to the northeast. It is about 50 acres in area with a maximum depth of 130 feet. It has failed to circulate since 1989, and has an anoxic hypolimnion, despite never being used for aquaculture. Complete mixing of the Twin City-South Lake is expected to become less frequent as the lake gets deeper. Additionally, inflowing groundwater is apt to be naturally cold, high in dissolved minerals, anoxic, and likely to sink to the lake bottom, thereby reinforcing any thermal and/or chemical density stratification, and any hypolimnetic oxygen deficit that already existed.

2.4   Basin Evolution and Lake Sediments

The mine pit side walls around the Twin City-South Lake are steep, barren of vegetation and extremely erodible. Storm runoff routinely causes large quantities of overburden soils surrounding the former mine pit to erode into the lake. Currently, the lake level is below the top of the bedrock stratum, and once it rises above this level, side wall erosion is expected to increase as wave action and inflowing groundwater attempt to establish a stable beach slope along in the lake's shore. Establishment of this stable angle of repose as the water rises will result in vast quantities of overburden being eroded into the lake and covering the current lake bottom. Eventually, the Twin City-South Lake basin will resemble a natural lake basin after erosional/depositional processes have established these stable slopes. This process may require decades or centuries, however. Until then, the lake will not be in chemical or biological equilibrium with its watershed.

Currently, natural sediment deposition of inorganic sands, silts and clays in the Twin City-South Lake is estimated by Axler *et al.* (1992) to be between 4 and 8 inches per year. This compares to only 4 inches, or less, of organic sediments that have been deposited in the lake as a result of aquacultural activities conducted there since 1988. The currently high rate of sedimentation in the Twin City-South Lake is expected to greatly increase after the lake surface rises above the top of the bedrock layer because of wave action against the overlying glacial till slope. The existing till slopes are very steep and the wave action will tend to cause instability resulting in increased sedimentation that will far exceed the current rate.

Thus, Twin City-South mine pit lake is an "embryonic" lake. That is, it has not yet reached an equilibrium condition in response to inputs from its surrounding watershed. As such, it does not have a natural "background" or "pre-aquaculture" water quality condition to which one can point as a logical lake restoration water quality goal. The water quality of this lake has been constantly changing since the mine pit began to fill with groundwater, and its ultimate steady-state condition is likely to be naturally meromictic because of its unique morphometry (size and shape). Also, Twin City-South Lake water quality varies frequently in response to the inputs of soil eroded from the pit walls by rainfall.

2.5   *Water Quality*

Water quality data have been collected from the Twin City-South mine pit lake since the start of MAI's aquaculture activates there in 1988. A detailed assessment of these data is provided by NRRI (Natural Resources Research Institute; Axler et al., 1992) and by MAI (1993). Generally, net pen aquaculture operations in the Twin City-South mine pit lake have resulted in enrichment of the lake due to releases of metabolic wastes (i.e., fish feces) and uneaten fish food that were dispersed into the lake water and/or sedimented onto the lake bottom. Annual waste phosphorus loads were approximately 7.7 g P/m^2 from 1990 through 1992.

Increased levels of phosphorus and nitrogen (ca. 115 µg P/L and 2,000 µg DIN/L) and periods of reduced dissolved oxygen (D.O.) concentrations in the water column have been documented since the aquaculture activities commenced. Increased deposition of organic matter on the lake bottom has also been observed beneath the net pens. Decomposition of this material exerts an oxygen demand on the overlying water column, which (if not satisfied) results in the release of phosphorus from the sediments.

Despite the addition of large amounts of nutrients and oxygen demanding organic matter, the Twin City-South mine pit lake has not shown highly degraded water quality conditions typical of eutrophied lakes. This is due to MAI's intensive aeration and circulation activities. Algal blooms have been prevented by vertical mixing of the water column, resulting in increased mixed layer depths and light limitation of algal growth. Mixing has also largely satisfied the sediment oxygen demand of the lake and prevented large releases of phosphorus from anoxic sediments. MAI aquacultural operations in the Twin City-South mine pit lake included the use of three banks of air diffusers and two large (8-foot diameter) submersible mixers. Two banks of diffusers, one each in the east and west regions of the lake basin, were operated at near-bottom depths to destratify the lake. A third set of air diffusers were operated at relatively shallow depth, just below the net pens, to oxidize relatively labile organic wastes falling out of the pens. The submersible mixers were positioned on opposite sides (north and south) of the lake and operated to create a basin-wide, counter clockwise current that constantly renewed the water in the net pens. Figures 1 and 2 illustrate, in concept form, the lake mixing system used by MAI in the Twin City-South mine pit lake.

During the 1989-1991 period, the estimated total oxygen demand of the Twin City-South mine pit lake peaked at about 5,800 mg D.O./m^2/day in August of 1991. The vast majority of this total, approximately 4,200 mg D.O./m^2/day, was related to fish respiration, while fecal matter and waste fish food were 1,400 and 200 mg D.O./m^2/day, respectively. Based on these figures and other data collected from the Twin City-South and Sherman mine pit lakes (NRRI, 1992; MAI, 1993), the peak 1993 daily oxygen demand of the Twin City-South mine pit lake is estimated to be about 500 kg D.O./day, following removal of the fish:

Sediment Oxygen Demand	20.6 kg O$_2$/day
Volumetric Water Oxygen Demand	480 kg O$_2$/day
Total Oxygen Demand	500.6 kg O$_2$/day

The dissolved oxygen data portrayed on Figure 3 demonstrate that, while it was operating, the MAI lake mixing system was adequate to destratify the Twin City-South mine pit lake during 1992. Mixing was temporarily suspended in late summer months because of concerns about the effects of possible nitrogen supersaturation on the fish, and about further epilimnetic oxygen depletion caused by the decomposition of suspended fish feces and waste fish food. The temperature and dissolved oxygen isopleth diagram shows the lake to be somewhat oxygen depleted near the lake bottom during the late summer periods.

Current lake water quality is vastly improved over previous years, as successive, algal blooms have removed phosphorus from the water column and subsequently settled into the lake's hypolimnion. The 1994 average summer (June-August) water quality conditions of the Twin City-South Mine Pit Lake were:

Total Phosphorus = 6.6 µg/L   Secchi disc transparency = 3.8 m
Chlorophyll a = 1.3 µg/L

These conditions are typical of a low-range mesotrophic or near-oligotrophic lake. Additionally, epilimnetic phosphorus concentrations continued to decline through the summer of 1993, to about 10 µg/L, whereas they rose to as high as 115 µg/L during aquacultural activities in 1992. The planned 1994-95 lake restoration activities -- Nutrient Inactivation with alum plus sodium aluminate, and subsequent Lake Bottom Sealing -- will further improve Twin City-South Lake water quality.

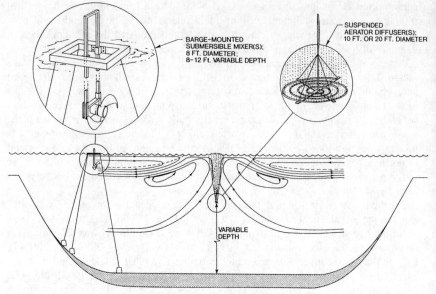

Figure 2
Aeration System -- Concept Plan
Twin City-South Mine Pit Lake

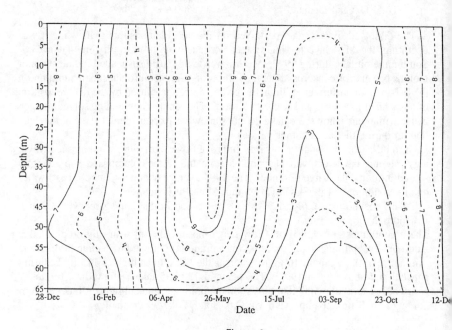

Figure 3
1992 Dissolved Oxygen Isopleths
Twin City-South Mine Pit Lake

3 LAKE RESTORATION PLAN

In preparing a Lake Restoration Plan for the Twin City-South mine pit lake (Barr, 1994), we selected restorative measures that have shown to be effective in improving lake water quality elsewhere. Our plan was predicated on the assumption that the Twin City-South mine pit lake is dimictic (twice mixing, annually), and that it would not soon become naturally meromictic (partially mixing) because of other factors unrelated to the previous aquacultural activities conducted there.

From many possible lake restorative measures, Nutrient Inactivation and Lake Bottom Sealing were selected as the techniques recommended to improve Twin City-South Lake water quality. These measures were chosen to control both the oxygen demand associated with the decomposing aquaculture waste, and the phosphorus being released by this material into the overlying water column.

First, a dual-purpose alum plus sodium aluminate treatment would be administered to the lake to remove phosphorus from the water column and to create an alum floc (i.e., aluminum hydroxide) blanket that would prevent the release of soluble phosphorus from anoxic lake sediments. The alum and sodium aluminate chemicals are typically applied to the lake in a slurry form through two separate sets of dosing nozzles on a boom towed behind a barge. The total required aluminum dose is achieved by alum and sodium aluminate additions in a 2:1 combination (volume basis). Such a dose will cause no change in the pH of the lake water.

Second, a confining layer of inert soil will be deposited over the alum floc blanket to limit the release of soluble oxygen demanding substances from the decomposing organic matter in the current lake sediments. This covering will consist of naturally occurring soils from the mine pit walls that will be eroded into the lake using a process termed hydraulicking -- basically, a barge-mounted water cannon shooting a high pressure stream of water against the mine pit walls. An approximately 8-inch thick layer of soil, which represents about one year's natural soil deposition in the lake basin will be used to seal the lake bottom. This points out the fact that, if left alone, the lake water quality would ultimately restore itself, without human intervention, to a level that would once again be in equilibrium with is watershed. The naturally high rate of sedimentation within the lake basin is estimated by NRRI to be between 4 and 8 inches per year. We estimated this rate of sedimentation would likely deposit enough inert soil over the current lake bottom (including aquacultural wastes) to essentially "restore" surface water quality to pre-aquaculture levels within several (1 to 2) years at no cost. Depending on how much phosphorus is added to the water column by the eroded soil, a second alum treatment may be performed, after hydraulicking, or the order of the planned lake restoration treatments may be reversed (i.e., make the alum plus sodium aluminate treatment after hydraulicking).

3.1 Nutrient Inactivation

To determine the dose of alum necessary to remove phosphorus from the water column of the lake, a standard dose rate determination test (Kennedy, 1978) was conducted. The results of these dose rate determinations suggest that an alum dose of 23.1 mg Al/L would be required to bring the Twin City-South Lake water pH down to the 6.0 level. [The lake water had a total alkalinity of 159 mg $CaCO_3$/L and an initial pH of 7.8, which means it would theoretically require an alum dose of

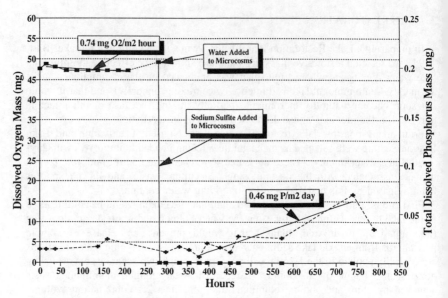

Figure 4. MAI Sediment Oxygen Depletion and Phosphorus Release Experiment:
Oxygen Depletion and Phosphorus Release Rates--8" Sand Cover Microcosm
@ 4 C--Without Alum

1994 End-of-Summer DO Profile   1994 End-of-Summer TP Profile

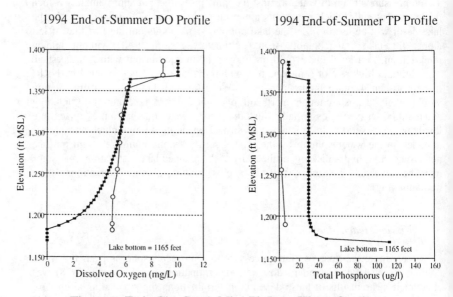

Figure 5. Twin City-South Mine Pit Lake Water Quality:
Model Predictions (-) versus Observed Conditions (o)

about 22 mg Al/L to reach the pH 6.0 level.] This treatment reduced the initial total phosphorus concentration of the sample from 0.129 mg P/L to only 0.007 mg P/L. Most of the phosphorus initially present in the sample, collected on February 17, 1993, was in the dissolved form ([TDP] = 0.122 mg P/L). Increasing the alum dose by as much as two times did not increase the observed phosphorus removal efficiencies, and reducing the dose to as little as 20 percent of the pH 6.0 dose did not appreciably lessen the phosphorus removal efficiencies. Therefore, a reduced dose alum treatment could potentially be as effective at removing phosphorus as a full dose treatment. However, in a lake as deep as the Twin City-South lake is (Z_{max} ≈225 feet, or 69 meters), alum treatments are more typically made on an areal dose rate basis than on a volumetric basis, especially when the secondary intent of the treatment is to provide an alum floc blanket over the lake sediments.

3.2   Lake Bottom Sealing

Several lake restoration projects have demonstrated the effectiveness of aluminum salts in suppressing the release of soluble phosphorus from anoxic sediments. Such alum or alum plus sodium aluminate treatments do not reduce the release of soluble oxygen-demanding substances into the overlying water column, however. Therefore, in order to maintain an aerobic hypolimnion in the Twin City-South mine pit lake, either the organic sediments must be sufficiently covered to limit the leakage of soluble oxygen-demanding substances into the overlying lake waters to a tolerable rate, or the hypolimnion must be aerated. Individually, the latter action is a problem treatment technique only, however, and would not "restore" water quality since the near-bottom waters would likely become anoxic again, once hypolimnetic aeration activities ceased.

To assess whether Lake Bottom Sealing with a layer of native mineral soil (primarily a fine sand with lesser amounts of silt and clay) was potentially feasible, sediment-water microcosm experiments were conducted utilizing Twin City-South mine pit lake waters and black, organic lake sediments (i.e., sapropel, according to Axler, et al. [1992]) collected from directly beneath the aquaculture net-pens there. This organic lake sediment was determined to have an ultimate (i.e., 16-day) biochemical oxygen demand that exceeded 40,000 mg kg^{-1}. The rates of oxygen depletion and phosphorus release by/from these sediments were then determined, under anaerobic conditions, after the addition of alum and/or varying amounts of soil to seal the organic lake sediment.

Sediment oxygen depletion rates were calculated over the first 208 hours of the experiment, and then the remaining dissolved oxygen in the microcosms was stripped out of solution by the addition of sodium sulfite (1.3X the stoichiometric requirement for complete oxygen removal). Following oxygen removal, the rate of sediment phosphorus release was gaged from total dissolved phosphorus concentration analyses of samples collected between 330 and 739 hours after the beginning of the experiments (a 17-day period). A portion of the results of these experiments is presented on Figure 4. For the "8-Inch Sand Cover" treatment that was intended to mimic conditions at the end of 1994 (one year after cessation of aquacultural activities in the lake), the sediment phosphorus release and oxygen consumption rates were found to be 0.46 mg P/m²day and 0.74 mg O₂/m² hour, respectively.

4 PHOSPHORUS MASS BALANCE AND DISSOLVED OXYGEN DEPLETION MODELING

Results of the water-sediment microcosm experiments from which sediment phosphorus release and oxygen consumption rates were calculated were used in phosphorus mass balance and dissolved oxygen depletion modeling of the Twin City-South mine pit lake. Areal hypolimnetic oxygen deficits (AHOD) were calculated according to the methods of Charlton (1980), modified to predict areal oxygen deficits for individual hypolimnetic strata and, therefore, hypolimnetic dissolved oxygen profiles, by substituting stratum volume/sediment surface area ratio for hypolimnetic thickness (Molot, et. al, 1992). Changes in phosphorus concentration were calculated assuming an apparent total phosphorus settling rate of 16 m yr^{-1}, according to Chapra (1975). Figure 5 portrays the results of our water quality modeling in comparison to late-summer conditions observed on September 29, 1994 by NRRI (C. Tikkanen, Personal Communication). Generally, observed water quality was better than we predicted, with little or no oxygen depletion or phosphorus release from hypolimnetic lake sediments. This disparity is probably due to the fact that burial of the organic lake sediments by naturally eroded clastic materials from the mine pit walls exceeded even our highest estimate of 8 inches (~20 cm) per year.

REFERENCES

Axler, R., C. Larson, C. Tikkanen, M. McDonald and G. Host. 1992. Limnological Assessment of Mine Pit Lakes for Aquaculture Use. University of Minnesota - Duluth; Natural Resources Research Institute. Technical Report NRRI/TR-92/03.

Barr (Barr Engineering Co.). 1994. Twin City-South Mine Pit Lake--Lake Restoration Plan. Prepared for Minnesota Aquafarms, Inc. for Submission to the Minnesota Pollution Control Agency.

Charlton, M.N. 1980. Hypolimnion Oxygen Consumption in Lakes: Discussion of Productivity and Morphometry Effects. Can. J. Fish. Aquat. Sci. 37: 1531-1539.

MAI (Minnesota Aquafarms, Inc.). 1993. Operations Reports, Unpublished Data, and Personal Communications from Mr. Dwight Wilcox, MAI Fisheries Biologist.

Molot, et al. 1992. Predicting End-of-Summer Oxygen Profiles in Stratified Lakes. Can. J. Fish. Aquat. Sci. 49: 2363-2372.

Tailings & Mine Waste'95 © 1995 Balkema, Rotterdam, ISBN 90 5410 526 7

Analysis of tailings and mine waste sediment transport in rivers

Robert K. Simons, Daryl B. Simons & Gilberto E. Canali
Simons & Associates, Inc., Fort Collins, Colo., USA

ABSTRACT: The transport of natural sediment, tailings and other mine wastes by rivers can be a significant issue in a variety of disciplines for new, existing and abandoned mines. Such issues may involve the development of appropriate tailings and water management plans, reclamation, or the mitigation of sediment and related issues. Either an understanding of the mechanism of tailings and natural sediment transport and deposition must be developed and accepted, or tailings and waste management plans must be developed and their effectiveness evaluated for the control of such materials to minimize environmental effects and to ensure long-term stability or appropriate reclamation. Recently Simons & Associates has been involved in a variety of mining projects where analysis of the transport and deposition of tailings is a significant issue for either current management plans or past mining operations requiring the estimation of tailings transport and deposition and potential mitigation to meet a variety of objectives. The tools used in such analyses include application of geomorphic analysis, engineering calculations, and physical-process computer models to develop the necessary understanding and to quantify riverine transport and depositional processes. This paper presents a discussion of the analysis approach as applied to the expansion of a mining project in Irian Jaya, Indonesia.

1 INTRODUCTION

An analysis of the transport and deposition of tailings in a river system has been conducted in relation to the expansion of an existing mine in Irian Jaya, a province of Indonesia located on the western half of the island of New Guinea.

The creation of the island of New Guinea and its associated mineral resources began about 14 million years ago when New Guinea arose from the sea as a result of the collision of two of the earth's great crustal plates. The Indo-Australian plate, which carries Australia, collided with and began moving under, or subducting, the huge Pacific plate which carries the Pacific Ocean. This collision and subduction deep under the surface pushed the ocean floor high above the sea creating an island with mountains in excess of 4,500 meters above sea level. This mountain building process continues today.

Deep beneath the mountains, the heat generated by plate collision and subduction turned rock into magma which was thrust upward into the mountains. On its way to the surface, this molten rock concentrated many important minerals such as copper, gold, and silver. Millions of years later, these minerals resulted in the establishment of one of the most significant mining operations in the world.

Before discussing the sediment transport analysis, it is first necessary to understand the physical processes at work in the area. The plate collision resulting in the uplift and mountain building process has created high mountains with very steep slopes. In the

AJKWA-OTOMONA-AGHAWAGON LONG. PROFILE

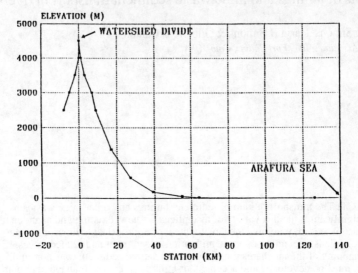

Figure 1

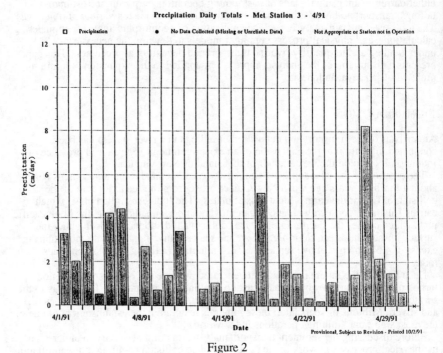

Figure 2

very short distance of approximately 125 kilometers, the mountains rise from sea level to over 4,500 meters. Most of the elevation gain in this region is concentrated in a very short distance of about 40 kilometers. Mountain and river slopes are very steep. Figure 1 displays a slice through the region showing the mountian peaks and profile of the river as it flows to the sea.

Being located close to the Equator on an island surrounded by the sea, available atmospheric moisture supplies are large, and therefore total annual rainfall is also large. Available data show that rainfall ranges from about 5 to 8 meters per year. Precipitation events occur almost every day of the year, and it is quite rare when there are more than a very few consecutive days without rain. An example of recently collected rainfall data is shown in Figure 2.

As a result of the large amount of rainfall on very steep slopes, deep canyons have eroded into the uplifted mountainsides. The volume of material eroded by the river system has been estimated to be approximately 440 billion cubic meters. This is the equivalent of about 730 billion metric tons. The canyons eroded in the mountainside often exceed 1 kilometer in depth.

As these materials erode from the steep mountain area, and as the water and sediment flow toward the sea, the slope and, hence, the energy to transport the sediment decreases resulting in sediment deposition and land building taking place. The land building processes result in a building upward and extending outward of the coastal plain. Based on the naturally eroded volume and the width of the coastal plain, this erosion raised the lowland area on the order of 500 meters in elevation.

This river, as it flows from the mountains to the sea, fits into the classic definition of the various zones of the idealized fluvial system as described in the scientific literature (Figure 3). These zones include the production zone where sediment supply is derived from erosion of the mountain watershed, the transfer zone where water and sediment flow - often times through braided channels over the alluvial plain towards the sea, and the deposition zone where much of the sediment not deposited in the development of the alluvial plain forms the classic multi-channel deltaic system approaching the coast.

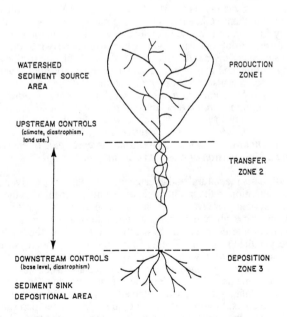

Figure 3   Idealized Fluvial System (after Schumm 1977)

The idealized system is now described as it relates specifically to this river system. The mountain zone is characterized by rapidly flowing rivers down steep slopes with tremendous turbulence and energy supplied by high precipitation. Velocities during high flow events can exceed 6 meters per second. This is the primary zone of sediment production. In a recent event, the river system was observed transporting large quantities of naturally eroded sediment including some particles in excess of 1 meter in diameter as well as large trees from landslide activity. In the alluvial plain, the steepness of the slope rapidly decreases along with the river's energy. It is here that braiding and meandering occur as the coarser fraction of the natural sediment load, including boulders and cobbles, begins to deposit. The larger size fractions settle in the upper reaches and smaller size fractions are transported down the river. The channels in the coastal lowlands are formed of sand, silt, and clay sized particles where much of the material eroded from the upper watershed is deposited in the lower zones in the land-building process.

Although this river system fits the classic zonal definition, it is important to note that fairly extreme conditions exist which govern the physical processes of this system. These extreme conditions include high rainfall, flood events with rapid changes in flow from low to high and back to low again in a matter of hours, high relief or change in elevation from the mountain tops to the sea, extremely short distance from the mountains to the sea resulting in extremely steep slopes, landslides, and earthquakes.

These extreme conditions result in a highly dynamic river system under natural conditions. The transitions from one zone to another that some rivers spread into thousands of kilometers are, in this river system, compressed into an incredibly short distance due to the existing topography.

These extreme conditions magnify the dynamics of these rivers. In other words, under natural condition, these rivers are highly dynamic, meaning they can change. Flood events and large scale natural erosion such as landslides with debris including large trees, makes these rivers highly dynamic. Old channels fill in or shift and new channels are created. This dynamic nature was well described by Schumm (1977)

"... an alluvial river generally is continually changing its position as a consequence of hydraulic forces acting on its bed and banks."

The evidence of channel dynamics is clearly shown by the existence of estuaries without rivers feeding into them. Once rivers fed these major estuaries, now they do not. These rivers naturally filled in and shifted or created new channels to the sea, creating other estuaries. As further proof, one can travel by boat from one river to the next without going to the sea, showing that the rivers have shifted back and forth over the coastal plain forming, filling, and abandoning channels and then forming new channels to the sea. This natural process is also discussed by Schumm (1977).

"The propensity for change is inherent in the coastal fluvial area... a characteristic of the coastal plain situation is that rivers can shift position by avulsion... Avulsion generally is in response to two factors: (1) channel aggradation due to progressive extension of a delta into the sea (increased length of the stream requires aggradation to maintain the gradient upstream), and (2) a shorter, steeper route to the sea that the river can adopt... The avulsion of rivers on alluvial plains and deltas is a normal process."

This dynamic condition is the natural and ever-changing state of these rivers.

Now let us turn our attention to the mining operation which is located in the upper mountains of the river system. Ore from the mine is transported to the mill located a short distance below the mine where the rock is ground into sand and smaller-sized particles. The minerals of interest are then physically separated from the finely-ground native rock in a process called flotation. The material produced by the physical separation process consisting of finely ground rock is called tailings.

Since tailings consist of finely ground native rock, they behave as do the naturally eroded sediments of the same particle size. Particles of this size range are generally transported through the mountain zone down to the alluvial plain and coastal zones. Thereafter, primarily the coarser portion of the material settles out and deposits in the lower two zones, while the finest portion of the tailings is transported towards the sea.

The deposition of these particle sizes is dictated by the ever decreasing slope as one approaches the sea. Deposition in this area has been, is, and will continue to be part of the on-going land building process.

In order to begin evaluating the effect of tailings it is first necessary to put in perspective the magnitude of the tailings as compared to the natural sediment erosion and deposition in this area. As previously mentioned, natural erosion forming deep canyons has eroded approximately 730 billion metric tons of material from the mountains with subsequent deposition in the lowlands on the order of 500 meters in depth. Natural river processes are currently being used to transport and selectively deposit tailings as the natural river dynamics dictate. Thus, the river places the tailings using the natural transport and depositional processes rather than trying to force the tailings to either go to or stay in locations where they would not go naturally.

Thus the natural river dynamic processes which include transport of sediment and tailings from the mountains and deposition in the lowlands will continue. As before the mine existed with natural erosion and deposition only, the river channels now with the mine in this highly dynamic environment will continue to change position, fill in and abandon old channels, and develop new channels as the process of erosion, deposition, and channel dynamics continues.

A number of tailings management alternatives have been developed and analyzed to determine what physically will become of the tailings and as a key initial step in developing an appropriate management strategy.

In order to develop an appropriate tailings management plan it is necessary to determine the physical fate of the tailings. This requires data including streamflow, channel geometry and sediment characteristics. At the beginning of this study very little data in these categories existed. Considerable effort was undertaken to obtain the necessary data. This paper presents the results of the data collection effort, various analyses of the data themselves, and then presents analysis using the data in sediment transport studies to better understand the physical processes of sediment transport as related to the production and transport of mill tailings.

2 DATA

Data pertinent to the analysis of sediment have been collected since 1990. For example, hydrological and meteorologic data have been collected on a continuous basis at numerous locations since about September 1990. In addition, other necessary data have been collected including riverine and estuarine channel geometry, bed material data, suspended sediment transport and other relevant sediment and tailings size distribution data. All of these data are necessary to develop an understanding of current conditions as mine production increases to higher rates commensurate with the increased amount of available ore. The pertinent data are now discussed in the following sections.

2.1 Channel Geometry Data

The channel geometry of a river is defined by its cross-sectional shape and by how steeply it slopes as it flows downstream. Both the channel cross section and slope vary from the upper portion of the watershed down to the sea. An example of the cross-sectional survey data is shown in Figure 4.

2.2 Sediment Data

2.2.1 Bed Material Data

In a similar way as the slope of these rivers vary from very steep to very flat as they flow from the mountains to the sea, the size of the bed material over which the water flows

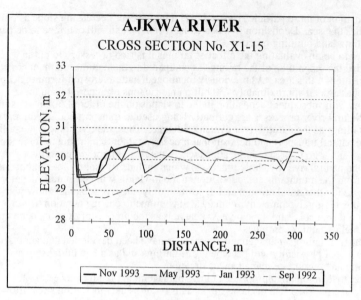

AJKWA RIVER
CROSS SECTION No. X1-15

Figure 4

varies from very coarse to very fine sized particles. In the mountains the bed of the river consists of large boulders, cobbles, gravel, and a very small amount of finer sizes such as silt and clay. The size of the bed material remains quite large, predominantly in the boulder category, from the upper watershed down to the foot of the mountains. Then, as the gradient of the river flattens and the velocity of flow and its energy decreases, the rivers capacity to transport the larger material also decreases and it can no longer transport boulders downstream. Below the mountain zone, the bed of the river consists primarily of cobbles with some generally smaller boulders, some gravel, and some finer silts and clays. An examples of the bed material of this type can be seen in the vicinity of the bridges. As the river continues to flatten as it approaches the coastal plain and delta, again the river's capacity to transport the coarser sizes (cobbles and gravel) further decreases. The bed of the river then consists primarily of sand with some silt and clay. A number of samples of the river bed material, primarily from the vicinity of the bridges downstream, were collected. These samples were then analyzed to determine the size distribution of the particles forming the bed of the river at each location.

The methodology of sample analysis to determine gradation differed depending on the size of bed material encountered. Where the sample was primarily cobble and boulder a photographic grid method was used. Where the sample consisted primarily of sand and gravel the gradation was determined by dry sieving. A combination of seiving and hydrometer was used where the sample had significant quantities of finer sized particles. Figure 5 presents an example of the bed material size distribution in the vicinity of the bridge.

2.2.2 Suspended Sediment Data

In order to determine the quantity of suspended sediment flowing down the river, suspended sediment samples were taken by the depth integrated technique using either a DH-48 or D-74 suspended sediment sampler. Samples were then analyzed to determine the concentration of suspended sediment by weight. The size distribution of suspended sediment particles for some samples was also determined. A portion of these samples

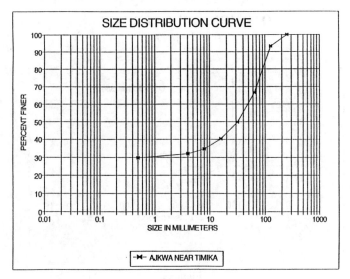

Figure 5

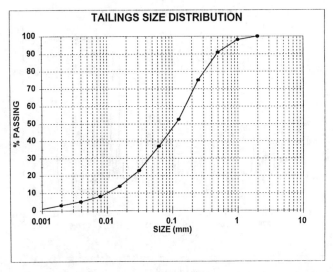

Figure 6

were analyzed using a particle counter, while others were analyzed using the more traditional methods of sieve and hydrometer. An example of these data both in terms of sediment concentration and size distribution are presented later in this paper.

2.2.3 *Tailings*

The size distribution of tailings, which consists of finely ground native rock, was also

97

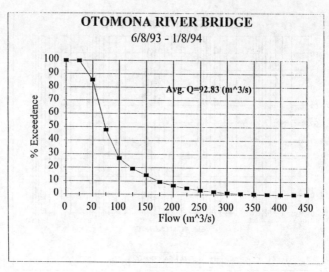

Figure 7

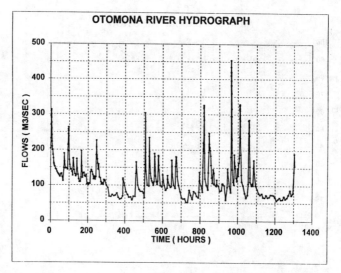

Figure 8

determined. Because of the relatively fine particle sizes found in the tailings, great care had to be taken in constructing the sieve analysis to ensure that the finer particles had been adequately separated from the coarser particles. Hydrometer analyses were conducted for the finer portion of the range of sizes. Figure 6 shows this tailings gradation.

2.3 *Flow Data*

Continuously recorded flow data are being collected at a number of gaging stations which have been established as part of the Long Term Environmental Monitoring Program (LTEMP). Some analysis of these data have been conducted to determine the magnitude of flood peaks for a range of return periods or frequencies. A flow duration curve and an example of the hydrograph are shown in Figures 7 and 8.

3 SEDIMENT TRANSPORT METHODOLOGY

Sediment transport depends upon a number of factors including magnitude and hydraulic characteristics of the flow. In turn, the hydraulic characteristics of the flow depend upon physical characteristics of the channel, including cross-sectional channel geometry, channel slope, and resistance to flow. Manning's equation is often used to determine hydraulic variables for a range of flow conditions based on the above characteristics and also based on the collected field data.

Manning's equation can be written as

$$(1) \qquad Q = \frac{1.49}{n} A R^{2/3} S^{1/2}$$

$$(2) \quad \text{or} \qquad V = \frac{1.49}{n} R^{2/3} S^{1/2}$$

where Q is the water discharge (cfs), n is Manning's resistance to flow parameter, A is the cross-sectional area (ft^2), R is the hydraulic radius (ft), S is the slope (ft/ft), and V is the velocity (ft/sec). This method correlates roughness and stream characteristics such as slope and hydraulic radius to velocity and the cross-sectional area flow and hence its depth and water surface elevation.

The hydraulic analysis which computes the velocity, hydraulic radius, and wetted perimeter for various stages and depths, provides the background information used in computing sediment transport capacity. The sediment transport capacity is composed of bed load and suspended load transport capacities. The bed load capacity is computed by the Meyer-Peter, Müller equation. This equation has been utilized in a wide variety of conditions and is expected to be applicable in this modeling effort. The suspended load capacity is computed by the Einstein method (Einstein, 1950; Meyer-Peter, Müller, 1948). Shields' criteria is used to determine the size of particles moved by the flow (Shields, 1936). The following procedure is used to determine sediment transport capacity.

3.1 *Bed Load*

The hydraulic parameters calculated above are used to determine sediment transport capacity. The first sediment transport parameter that is determined is the tractive force or boundary shear stress. The boundary shear stress acting on a grain can be determined by:

$$(3) \qquad \tau = \frac{1}{8} \rho f v^2$$

where τ is the boundary shear stress, ρ is the density of water, and v is the velocity. The critical shear force for particle movement is determined by the following Shields' criteria (refer to Shields, 1936, and the Task Committee on Preparation of Sediment Manual, 1966).

(4) $$\tau_c = \tau_* \gamma (S_s - 1) d_{si}$$

where τ_c is the critical shear stress (lbs/ft²), τ_* is the Shields' parameter (dimensionless), S_s is the specific gravity of the sediment (usually 2.65), and d_{si} is the sediment size (geometric mean for the fraction) (ft). If τ is greater than τ_c there is sediment movement for this size class. The bed load transport is calculated with the following Meyer-Peter, Müller formula:

(5) $$q_{bi} = \frac{12.85}{\sqrt{\rho}} (\tau - \tau_c)^{1.5}$$

where q_{bi} is the bed load transport rate pounds/sec.ft, ρ is the density of water (lbs-sec²/ft⁴), and τ and τ_c are as previously defined.

3.2 Suspended Load

The suspended load is determined using the Einstein method also known as the Einstein integral as follows:

(6) $$q_{si} = \frac{q_{bi}}{11.6} \frac{S_a^{w-1}}{(1-S_a)^w} \left[\left(\frac{V}{U_*} + 2.5 \right) J_1 + 2.5 J_2 \right]$$

where w is a dimensionless parameter that relates settling velocity of the sediment in water to the shear velocity

(7) $$w = \frac{V_s}{k U_*}$$

V_s is the settling velocity (ft/sec), k is von Kármán's number, taken as 0.40, and U_* is the shear velocity (ft/sec). Settling velocities are a function of particle size and water properties. They are calculated as:

(8) $$V_s = \frac{2.9517 d_{si}^2}{v} \quad \text{when} \quad d_{si} < 0.0002 \text{ ft}$$

(9) or $$V_s = \frac{(36.064 d_{si}^3 + 36v)}{d_{si}} \quad \text{when} \quad d_{si} \geq 0.0002 \text{ ft}$$

where v is the dynamic viscosity of the water (ft²/sec), and V_s and d_{si} are as previously defined. The shear velocity is given by the equation:

(10) $$U_* = (\gamma RS/\rho)^{1/2} \quad (\text{ft/sec})$$

S_a is the dimensionless parameter that relates flow depth to sediment size as:

(11) $$S_a = 2d_{si}/R$$

100

where d_{si} and R are as previously defined.

The terms J_1 and J_2 are integrals resulting from integration of the equation describing the vertical concentration of the sediment in the flow. The first integral J_1 is given as:

$$(12) \qquad J_1 = \int_{S_a}^{1} \frac{(1-\sigma)}{\sigma} d\sigma$$

where σ is a relative position, and

$$(13) \qquad \sigma = \varepsilon/\gamma \ (ft^2/lbs)$$

And ε is the distance of the sediment particle above the stream bed in the flow. The other integral is similar and is given as:

$$(14) \qquad J_2 = \int_{S_a}^{1} \ln\sigma \frac{(1-\sigma)^w}{\sigma} d\sigma$$

These two integrals can be evaluated by successive integrations of a power series expansion, or by graphic inspection (Simons and Sentürk, 1992).

These computed values are then included in the suspended sediment equation. A size class will be transported in suspension if it satisfies the following condition:

$$(15) \qquad \left[\left(\frac{V}{U_*}+2.5\right)J_1 + 2.5J_2\right] > 0$$

Otherwise, the size may be transported only as bed load. The value for total sediment transport capacity for each size fraction can be computed for all the cross sections.

$$(16) \qquad Q_i = (q_{bi} + q_{si})WI$$

where Q_i is the total sediment transport capacity (bed load plus suspended load) for the size fraction i (lbs/sec), W is the cross section width for the stage (ft^2), and I is the size fraction for the size i (%). The total sediment transport capacity of the discharge Q is then computed by the integration of the transport capacity of each size fraction.

$$(17) \qquad Q_s = \sum_{i=1}^{N} Q_{si}$$

where Q_s is the total transport capacity, and N is the number of size fractions.

4 SEDIMENT TRANSPORT MODELING

Based on data collected in the past few years and using the methodology described in the previous section, a sediment transport model was used to analyze sediment transport in the river system. One of the primary objectives of using the model was to determine the fate of sediment being transported down these rivers and assist in developing an appropriate tailings management plan.

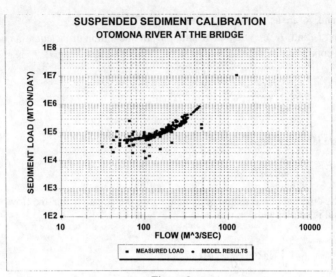

Figure 9

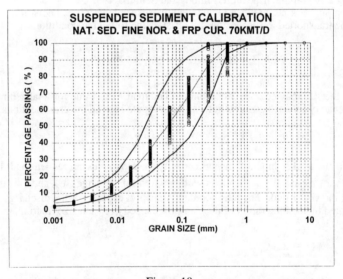

Figure 10

The sediment transport model was set up based on the available data as previously described. The upstream supply of sediment was based primarily on the rate of tailings production and natural sediment transport. A representative period for flow and sediment supply data was selected to test the model. Using the channel geometry and upstream sediment supply data as input, the model was used to compute sediment transport down the river system as it changes as the river gradient flattens in the downstream direction. The results of the model were compared with observed data collected about 40 miles downstream of the point of tailings discharge into the river. Figure 9 shows the measured suspended sediment transport data compared with computed values plotted as a function of flow. As can be seen in this figure, the computed results compare favorably with the measured values. Figure 10 shows the computed size distribution for suspended sediment compared to the measured data based on the laboratory analysis of the data. Again, the computed values produced by the model compare favorably with the data.

5 DISCUSSION OF RESULTS

With the analysis comparing the computer model results to the data showing good agreement, the model was used to evaluate various scenarios including increased tailings production as well as alternative tailings management options. In other words, the supply of tailings was increased in the model to reflect expanded production projecting the areas of tailings deposition under these conditions. The model was also used to test various sediment and river control options to evaluate the effectiveness of a range of potential tailings and river management plans. These evaluations were presented to mining managers for a decision on an appropriate approach. This information is then submitted for approval to the various appropriate government agencies who regulate mining activity and environmental impacts. The model has been demonstrated to be a useful tool in the quantification and evaluation of the sediment transport issues in developing an appropriate tailings management plan.

REFERENCES

Einstein, H.A., 1950. "The Bed-Load Function for Sediment Transportation in Open Channel Flows," USDA, Sol Conservation Service, Technical Bulletin No. 1026.
Meyer-Peter, E., and Müller, R., 1948. "Formulas for Bed-Load Transport," *Proceedings*, Third Meeting of Intern. Assoc. Hydr. Res., Stockholm, pp. 39-64.
Schumm, S.A., 1977. The Fluvial System. John Wiley & Sons, New York.
Shields, I.A., 1936. "Application of Similarity Principles and Turbulence Research to Bed-Load Movement," A translation from German by W.P. Ott and J.C. van Vchelin, U.S. Soil Conserv. Service Coop. Lab., California Inst. Technology, Pasadena, 21 p.
Simons, D.B., and Sentürk, F., 1992. Sediment Transport Technology - Water and Sediment Dynamics. Water Resources Publications, Littleton, Colorado.
Task Committee on Preparation of Sedimentation Manual, 1966. "Sediment Transportation Mechanics: Initiation of Motion," J. of the Hydraulics Div., ASCE, Vol. 92, HY2, MArch, pp. 291-314.

Stream restoration and mine waste mangement along the upper Clark Fork River

C.T.Stilwell
Atlantic Richfield Company, Anaconda, Mont., USA

1 ABSTRACT

Historical mining activity has resulted in mining waste, or tailings, being deposited in and along the floodplain of several streams in the upper Clark Fork River basin in Southwestern Montana. These tailings, along with increased sediment and contaminant loading from historical and present agricultural practices and domestic sewage discharges, have impacted the streams and riparian habitat of the floodplain. The tailings, which are sulfidic, acid-generating sand, silts, and clays, contain elevated levels of arsenic, cadmium, copper, lead and zinc. They were deposited in the floodplain during several large flood events.

There are several reclamation designs and techniques being considered for use in the reclamation of tailings-impacted streambanks and floodplains of the streams in the upper Clark Fork River basin. This reclamation is currently being investigated under the Clark Fork River and Butte/Silver Bow sites designated under CERCLA (Superfund). Four of the basic designs which may be used as part of the reclamation efforts associated with riparian corridors are: in situ lime treatment; selective removal; combined selective removal/in situ treatment; and total removal.

The designs are all similar with respect to their ultimate goal, which is to reclaim a tailing-impacted stream reach to a sustainable natural riparian setting and a viable cold water fishery. Specific design objectives which must be met to achieve this goal are:

a. Reduce significant dissolved metal and sediment impact introduced to stream during precipitation-induced run off from floodplain;

b. Reduce significant sediment and metal impact introduced to stream by streambank erosion from fluvial processes (stream meander) and flood events;

c. Reduce significant metal input to groundwater due to leaching from the streambank tailings.

If these objectives are met, the significant exposure pathways will be reduced/eliminated to allow the affected terrestrial and aquatic ecosystems to be restored.

All designs include manipulating the fluvially-deposited tailings (by in-situ lime-amending and/or removing them), stabilizing the streambanks, and revegetating the floodplain and streambank areas. Similarities in implementing the designs include employment of conventional earth-moving and other construction equipment. The primary distinction between the various designs are the type and degree of effort required in manipulating the fluvially-deposited tailings. Case studies of projects implemented in the upper Clark Fork River basin are available to illustrate and compare each of the four designs.

1.1 *In-situ lime treatment*

The in-situ treatment design is centered primarily on the use of a treatment technology in which lime (calcium carbonate [CaCO] and calcium oxide [CaO]) is mixed with the mine tailings, then revegetated. The treatment is implemented by first mixing in lime to neutralize acid-generating potential of the tailings. The tailings are tilled using conventional agricultural tilling plows or more specialized deep tilling plows to thoroughly mix the lime, tailings and available native soil together. The physical soil matrix is also improved during tilling, thus facilitating establishment of vegetation.

Once the lime is mixed in, the tailings and soil is mulched, fertilized and revegetated. Reseeding is performed by drill seeding in a mixture of seeds in late fall or early spring and allowing the plants to naturally establish during the growing season. The mixture of seeds used in revegetation have been lab- and field-tested to be salt tolerant and drought tolerant. The seed mix contains some native species for long-term stand viability and some introduced species for rapid stage establishment. Vegetation stands need to be established quickly to minimize erosion and invasion of weeds.

The lime amendments and revegetation techniques and technology used is consistent with those studied under the Streambank Treatment and Reclamation Study (STARS) performed by Montana State University's Reclamation Research Unit. Also, barren erosive streambanks are pulled back to a 3:1 slope, lime-treated (if necessary), reseeded, and covered with biodegradable erosion-control fabric. In addition, native willows are sprig along the the streambank for eventual increased streambank stability. The case study for this option is the Clark Fork River Demonstration Project (or the Governor's Project).

1.2 *Selective removal*

The second design, selective removal, requires the removal of the primary metal-containing and acid-generating tailings to an off-site repository, lime-amending the remaining native soil surface, and revegetating. No metal concentration removal criterion (chemical criteria) is used, therefore, some elevated concentrations of metals are expected to remain in the native soils left in place. The theory is once the primary source of metals is removed and the remaining soil is revegetated, metals will not be available to be mobilized to impact the surface or groundwater. Also, no soil replacement is done. An important aspect in facilitating the application of this removal technique is if the tailings are laid into the floodplain in a discrete and visually discernible layer. This is important because determination of where tailings are to be removed can be done using visual criteria, instead of chemical analysis. Another important consideration in not using a chemical-based removal criteria is in removing tailings which are interspersed with healthy, viable vegetated areas. Tailings can be "surgically removed" and well-vegetated areas can be left undisturbed. Also, barren, erosive streambanks are removed, seeded, and in some cases, temporarily protected by employing erosion-control fabric. Willows are spriged along the streambank for increased stabilization. The case study for this design option is the Silver Bow Creek Demonstration Project II near Opportunity, Montana.

1.3 *Total removal*

Design option 3, total removal, entails removal and off-site disposal of tailings and metal-containing soils, re-constructing floodplain area where needed, and revegetation. The primary differences between "total" removal and "selective" removal is the degree of removal in areas containing elevated concentrations of metals. Total removal presumes most or all material with elevated metal concentrations must be removed.

Establishing a meaningful chemical-based removal criteria, then meeting this criteria in implementing the design is required. Total removal may also require soil replacement to reconstruct the floodplain prior to revegetation. The case study for this design is the Mill-Willow Bypass associated with the Warm Spring Ponds near Anaconda, Montana.

1.4 *Combined selective removal/in-situ treatment*

The fourth and last design combines STARS-type lime amendment and revegetation technology with partial removal and on-site relocation of tailings. The criterion for removing tailings is based on where in-situ treatment may be less effective in stabilizing/isolating the effects of the streambank tailings. The reason for removing tailings under these conditions instead of lime-amending all tailings-impacted areas was evidence that STARS-type lime amendment/revegetation techniques are less effective where groundwater is near the root zone or a constrained stream channel allows higher erosive effects. In most cases, the only areas fitting this criterion were directly adjacent to the creek.

The construction of this design requires excavating fluvially deposited sediments and mine tailings within a corridor along Silver Bow Creek, relocating those materials along the margins of the 100-year floodplain of the creek, and bringing in clean borrow material to reconstruct the excavated areas. The relocated materials will be placed in flood protected areas, lime-treated, and revegetated. Soils located outside of the excavated corridor but within the 100-year floodplain will be amended with lime. When excavation, backfilling, and lime incorporation tasks are completed, the site will be subjected to a final grading to drain towards Silver Bow Creek. Revegetation will consist of application of fertilizer, drill-sedded with an assembly of grass species and willows spriged into streambanks. The case study for this design is the Silver Bow Creek Demonstration Project I, near Rocker, Montana.

The following criterion is used to compare the designs: effectiveness in achieving the design goal and objectives; feasibility of implementing design; and cost. Information for addressing these criteria, as well as other relevant environmental, construction, and regulatory issues pertaining to reclamation of mine tailings in a riparian/stream setting will be presented. This information is available from construction and monitoring of the respective case studies for each of the four designs. This information will also be the basis for development of the Feasibility Studies and remedial discussions for the Superfund remediation on Silver Bow Creek and the Clark Fork River.

Prediction of tailings effluent flows

D. E. Welch & L. C. Botham
Golder Associates Ltd, Mississauga, Ont., Canada

J. M. Johnson
Golder Associates Inc., Denver, Colo., USA

ABSTRACT: A tailings basin has to be able to operate over a large range of operating and hydrological conditions. This paper describes a simple, computerized spreadsheet program called WATBAL which can predict flows and carry out sensitivity analyses. The input data can be entered into the model in simple, easily recognizable terms which facilitates operation.

The prediction of flow in a tailings basin is extremely important from an environmental point of view. Environmental impact, and of course costs, are directly related to the quantity of effluent that has to be treated and discharged to the environment. An important objective is therefore to minimize flows. A little effort in addressing the parameters that affect flows can have a beneficial impact such as watershed area, discharge slurry density, recirculation to the mill and miscellaneous inflows.

In the paper, a hypothetical example is worked through for a 3000 t/day gold mill for climatic conditions in Northern Ontario, Canada. Sensitivity analyses are carried out and best and worst case operating scenarios are established which show that the net annual flow for a reasonable set of worst case parameters can be easily three or four times average flow. The example clearly demonstrates that there is no such thing as a unique water balance for a tailings basin and that water management facilities must be designed to handle a wide range of flows.

1 INTRODUCTION

Tailings basin management is primarily a water management problem. It is relatively easy to effectively contain tailings solids but the effluent has to be safely passed through the system in a never ending stream either to the environment or recycled to the mill. In dry climates, tailings water has to be conserved for reuse. The prediction of flows is therefore extremely important especially from an environmental point of view, both during the operating period and after closure. Cost and environmental impacts are directly related to the quantity of effluent that has to be discharged to the environment. An important objective is therefore to minimize flows both in dry and wet climates.

There is no such thing as an unique water balance for a tailings basin because the input parameters can vary from day to day, month to month and even yearly.

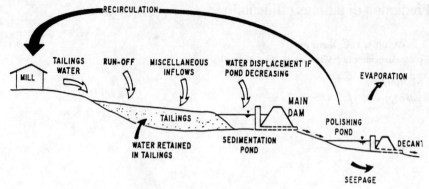

Figure 1. Elements of a tailings basin water balance.

Average, base case conditions may not even be the norm. Precipitation is the greatest variable, discharge slurry density can change, evaporation is never constant, recirculation to the mill is frequently a moving target and even the milling rate can change. Optimal effluent management can be achieved by:

1. keeping the watershed as small as possible to minimize run-off,
2. providing large ponds to promote evaporation or in the case of dry climates small ponds to inhibit evaporation,
3. maintaining a high slurry density to minimize the process water, and
4. maximizing the recirculation back to the mill.

If these conditions are optimized it is possible to develop a facility with minimal discharge to the environment and low treatment costs.

This paper describes a computerized water balance model called WATBAL which has been developed by Golder Associates and used on many tailings basins in Canada over the years to predict flows, size treatment plants, predict pond levels and to determine schedules for dam raising. It is a spreadsheet based program which only requires 50 kBytes of disk storage space in addition to the disk space requirements for the software. The program will run efficiently with the minimum computer system memory (RAM) specified for the software. The WATBAL model has been previously discussed in other fora (Welch and Firlotte, 1989) and (Welch, Botham and Bronkhurst, 1992).

WATBAL is merely a mathematical tool which adds and subtracts inflows and losses to a system. The results from the output are only as good as the numbers that are put into it, however, it has proven to be an effective program. The user must ensure, however, that the assumptions and input data are appropriate for the intended application. The advice of a hydrologist should be sought to assist with the interpretation of hydrological records, the estimation of run-off and the routing of storm flows. The program is a simplified model of simulated field conditions for a small watershed operating under specified hydrological conditions. Storm flows have to be handled separately.

2 INPUT DATA

A tailings basin must be designed to operate within a range of operating and hydrological conditions. All inflows and losses in a basin must be identified and accounted for in the water balance. The elements of a typical water balance are shown on Figure 1. Every effort should be made to minimize inflows to the basin while maximizing the re-use of tailings water in the milling process.

Prior to modelling the basin, a realistic set of design parameters has to be established for a 'base case' which represents normal (average) operating and hydrological conditions along with, 'best' and 'worst' case scenarios to establish the boundary conditions on which to base the design of pumping system, pipelines and treatment plant.

For this paper, input data for a hypothetical gold mill operating in a relatively wet climate with severe winters is used. The base case data for such a mill is as follows:

Tailings production	3000 t/day (metric)
(averaged over 365 days/yr.)	
Operating period	12 months/year
Discharge slurry density	45% solids by weight
Miscellaneous inflows	none
Precipitation (average annual)	850 mm/yr. (33.5 inches/yr.)
Total watershed area	2.5 km^2
Area of tailings and ponds	1.5 km^2
% run-off (run-off factor)	
- virgin land	75% of precipitation
- tailings and ponds	100% of precipitation
Run-off period	8 months (April to November)
Specific gravity of tailings	2.70
Assumed void ratio of deposited tailings	1.0
Estimated dry density of tailings	1.35 t/m^3 (84.3 lb/ft.^{3})
Saturated water content of tailings	37% (wt. water/dry wt. tails)
Estimated basin seepage	2500 m^3/month (15 USgpm)
Lake evaporation from ponds	
and wetted tailings surface)	450 mm/yr. (17.7 inches/yr.)
Area of pond and wetted tailings beach	0.4 km^2
Recirculation to the mill	90% of discharge water
Decant strategy	6 months (mid May to mid November)

The input data can be entered into the model in simple, easily recognizable terms which facilitates changes and enables sensitivity analyses to be carried out to investigate conditions within a range of input parameters. The input data, as they are entered into the program for the base case, are shown on the upper part of Table 1 and the detailed computations on the lower part of Table 1.

3 LOGIC

WATBAL compares inflows and losses on a monthly basis taking into account precipitation, tailings transport water, and miscellaneous inflows such as mine water,

111

Table 1. Example Water Balance Base Case Conditions

Input Parameters

	Units	Value	Jan	Feb	Mar	Apr	May	June	July	Aug	Sept	Oct	Nov	Dec	Total
Starting Month	(no.)	1													
Mill Production	(t/day)	3000													
Operating Period	logic		1	1	1	1	1	1	1	1	1	1	1	1	
Solids (by Weight) in Discharge	(%)	45													
Miscellaneous Inflows	(cu m/mo)		0.0	0.0	0.0	0.0	0.0	0.0	0.0	0.0	0.0	0.0	0.0	0.0	0.0
Precipitation	(mm/mo.)		50.0	45.0	55.0	60.0	75.0	80.0	85.0	90.0	100.0	75.0	80.0	55.0	850.0
Area Virgin Land in Basin	(ha)	150													
Runoff Coefficient	(%)	75													
Area of Tailings and Ponds	(ha)	100													
Runoff Coefficient	(%)	100													
Period of Runoff	logic		0	0	0	1	1	1	1	1	1	1	1	0	
Water retained in Tailings (water content)	(%)	37													
Estimated Seepage Losses	(cu m/mo)	2500													
Evaporation	(mm/mo.)		0.0	0.0	0.0	0.0	60.0	105.0	110.0	90.0	55.0	25.0	5.0	0.0	450.0
Area of Ponds and Wetted Tailings	(ha)	30													
Recirculation	(%/mo.)	90	0.0	0.0	0.0	0.0	8.3	16.7	16.7	16.7	16.7	16.7	8.2	0	
Decant Strategy (portion of total net inflow)															
Initial Water Volume in Ponds	(u m)	500000													

Computations

	Inflows				Losses			Accumulation					
Month	Tailings Water	Misc. Inflows	Runoff	Total	Retained in Tailings	Seepage Losses	Pond Evap.	Recirc.	Total	Net Inflow	Decant	Net Change	Accum. Volume
1	2	3	4	5	6	7	8	9	10	11	12	13	14
Initial													500000
Jan	111540	0	0	111540	33744	2500	0	100386	136630	-25090	0	-25090	474910
Feb	111540	0	0	111540	33766	2500	0	100386	136652	-25112	0	-25112	449798
Mar	111540	0	0	111540	33744	2500	0	100386	136630	-25090	0	-25090	424708
Apr	111540	0	563125	674665	33766	2500	18000	100386	136652	538013	0	538013	962721
May	111540	0	159375	270915	33766	2500	18000	100386	154652	116263	113706	2557	965278
Jun	111540	0	170000	281540	33766	2500	31500	100386	168152	113388	228781	-115394	849884
Jul	111540	0	180625	292165	33766	2500	33000	100386	169652	122513	228781	-106269	743616
Aug	111540	0	191250	302790	33766	2500	27000	100386	163652	139138	228781	-89644	653972
SAep	111540	0	212500	324040	33766	2500	16500	100386	153152	170888	228781	-57894	596079
Oct	111540	0	159375	270915	33766	2500	7500	100386	144152	126763	228781	-102019	494060
Nov	111540	0	170000	281540	33766	2500	1500	100386	138152	143388	112336	31052	525112
Dec	111540	0	0	111540	33766	2500	0	100386	136652	-25112	0	-25112	500000
Total	1338480	0	1806250	3144730	405150	30000	135000	1204632	1774782	1369948	1369948	0	500000

less losses to the system including evaporation, water retained in tailings, seepage and recirculation to the mill. An appropriate decant strategy can be applied to the model which then calculates the net change and accumulated volume on a monthly basis. The sensitivity of the system can be investigated by simply varying the input parameters.

With reference to the detailed output on Table 1, the balance is set up as follows:

Inflows
   Column 2 - Tailings water
   Column 3 - Mine water (miscellaneous inflows)
   Column 4 - Run-off
   Column 5 - Inflows summation

Losses
   Column 6 - Water retained in tailings
   Column 7 - Seepage
   Column 8 - Pond evaporation
   Column 9 - Recirculation
   Column 10 - Losses summation

Accumulation
   Column 11 - Net inflows (Col. 5 minus Col. 10)
   Column 12 - Rate of decantation
   Column 13 - Net monthly charge
   Column 14 - Accumulated pond volume

The tailings water (**Column 2**) is computed from the percent solids (by weight) in the tailings slurry by solving for W_w in the following formula:

$$\% \ solids \ = \left(\frac{W_s}{W_s + W_w} \right) 100$$

where: W_s = dry weight of solids
       W_w = weight of water

Miscellaneous inflows such as mine water (**Column 3**) are sometimes put into the tailings pond but this practice is becoming less common particularly if the effluent discharge is to be minimized. It is normally better to handle mine water separately because it usually has different holding and treatment requirements.

Run-off is computed in **Column 4**. The model can account for run-off from two different portions of a watershed; virgin land around the tailings which drains into the pond and the tailings and ponds. Run-off from the virgin land is computed as a percentage of precipitation, which for a rocky slope surrounding a tailings area in Northern Ontario could be in the order of 70 to 75%. This value takes into account evaporation and transpiration. All of the precipitation that falls on the tailings surface and ponds is assumed to enter the ponds therefore the run-off factor is 100%. If the tailings are not completely saturated then some of this precipitation could be attenuated in the tailings before reaching the ponds but eventually it will get there. Evaporation from the pond is accounted for separately in model.

In the winter the run-off is greatly reduced because snow is held until the spring thaw. The model can accommodate this by providing for zero run-off during the winter months and allowing the accumulated precipitation to enter the pond in the spring. This is a conservative but not unrealistic assumption.

The hydrological data are normally available from the hydrological services in a monthly format which is suitable for direct use in the water balance. For our hypothetical case, the monthly distribution used is typical for Northern Ontario.

Column 5 is the summation of the inflows.

Water is retained in the pore spaces of the tailings (**Column 6**) and is entered into the program by water content in the traditional soil mechanics sense which is the ratio of the weight of water over the dry weight of tailings. If the tailings are saturated then the water content can be simply calculated as follows:

$$w = \left(\frac{\rho_w}{\rho_d} - \frac{1}{G_s} \right) 100$$

or

$$w = \left(\frac{e}{G_s} \right) 100$$

where:

w = saturated water content (%)
ρ_d = dry density of tailings
G_s = tailings specific gravity
ρ_w = density of water
e = void ratio

If a portion of a tailings mass is partially drained then the water content has to be reduced by estimation.

Unrecoverable seepage (**Column 7**) may not be accurately known unless a full scale geological and hydrogeological assessment has been made of the basin. If this has not been done then the seepage has to be estimated. In any event, seepage does not normally represent a significant volume with respect to the balance although it can be an environmental problem.

Evapo-transpiration is taken into account in the run-off factor that is applied to the precipitation falling on the watershed surrounding the tailings. However, once the run-off reaches the ponds, water can be lost through evaporation. Lake evaporation (as opposed to pan evaporation) is therefore applied to the ponds and wetted surface of the tailings (**Column 8**). Evaporation data is normally obtained from the same sources as precipitation, also in a monthly format. Once again, the distribution that has been used in our hypothetical example is typical for Northern Ontario.

Water, that is recirculated from the tailings pond for re-use in the mill (**Column 9**), is entered into the model as a percentage of the tailings discharge water. All the water that enters a mill is either recycled within the mill or is discharged as tailings water.

Column 10 is the summation of the losses.

The "Net Inflow" (**Column 11**) is the excess water that accumulates in a tailings pond. It is the difference between Columns 5 and 10. It is the volume that has to be decanted and treated on an annual basis if the pond size is to remain constant. If the pond size decreases, by tailings displacing water, then the effective volume of water, which has to be decanted, will increase accordingly. This is discussed later in the paper.

The amount of water decanted each month is entered into the program as a percentage of the total net inflow (**Column 12**). It is controlled by design and can vary from month to month, and even be zero. However, if the system is to remain in balance, on an annual basis, then the total water decanted must be equal to the net annual inflow. **Column 13** gives the net monthly change, and **Column 14** the accumulated volume on a monthly basis.

4 SENSITIVITY ANALYSES

A little effort in addressing the parameters affecting the operation of a tailings basin can often have a beneficial impact on development and operating costs. The water balance can be used to quickly investigate the effect that the various parameters will have on the volume of water that has to be treated and discharged to the environment;

1. diversion or reduction in watershed area,
2. discharge slurry density,
3. recirculation to the mill,
4. precipitation and evaporation,
5. miscellaneous inflows such as mine water,
6. water retained in the tailings (pore water),
7. pumping out existing water in a basin prior to start-up, and
8. seepage.

For example, if the slurry density is increased from 30% to 50% solids, the process water is reduced by more than half. Runoff is of course a function of watershed size and diverting part of a watershed may be an economical trade-off against the management of a larger volume of water. Evaporation is a function of pond area and wetted tailings surface. Large shallow ponds encourage evaporation. A high recirculation rate reduces the need to introduce clean make-up water to the system.

As is mentioned above, a tailings basin has to be able to operate within a range of conditions with safety measures and back-up strategies in the event of unusual conditions occurring. A sensitivity analysis can be used to help define that range and to establish reasonable worst and best case scenarios. For our example a reasonable set of best and worst case parameters could arguably be:

Parameter	Best Case	Average Base Case	Worst Case
Tailings production	3000	3000	3000
Slurry density (%)	60	45	25
Miscellaneous inflows			
- Mine water (m³/day)	0	0	1500 (275 USgpm)
Precipitation	-25%	average	+25%
Recirculation (%)	100	90	30

5 RESULTS

As is mentioned above, the computer input and output for the 'base case' are given on Table 1. The model computes flows on a monthly basis and the annual balance, which is summarized below, is merely the total line on Table 1.

INFLOWS (Mm^3)

Tailings water	1.34
Miscellaneous inflows	0
Runoff	1.80
TOTAL	3.14
LOSSES	
Retained in tails	0.40
Seepage	0.03
Evaporation	0.14
Recirculation	1.20
TOTAL	1.77
NET ANNUAL INFLOW	1.37

The results of sensitivity analyses on significant parameters are given on Table 2. The analyses are carried out by varying the input parameter being studied and keeping all of the other parameters constant at the base case.

All the parameters investigated on Table 2 have a significant impact on the balance. If the annual precipitation increases by 25% then the net annual inflow increases by 36%.

Table 2. Results of Sensitivity Analyses

Parameter	Range	Variation in Net Annual Inflow	
		Mm^3	% Change
Precipitation	+25%	1.82	+36
	* Average	1.37 *	-
	-25%	0.92	-30
Slurry Density (% solids)	20	1.67	+66
	30	1.49	+26
	* 45	1.37 *	-
	60	1.31	-13
Recirculation (%)	0	2.57	+88
	30	2.17	+58
	60	1.77	+29
	* 90	1.37 *	-
	100	1.24	-10
Mine Water (m^3/day)	* 0	1.37 *	-
	500	1.55	+13
	1500	1.92	+40
	2500	2.28	+66
	3000	2.46	+80

Notes:
1. * denotes base case condition.
2. The sensitivity analyses are carried out by varying the parameter being investigated and keeping all the other parameters constant at the base case.

If the slurry density were allowed to decrease from 45% solids to say 20% solids then the net annual inflow would increase by 66%.
Recirculation has a major impact. A drop to "0" increases a net inflow by 88%.
Putting the mine water into the tailings pond has a major compact. At 3000 Mm³/day the net annual inflow would increase by 80%.
As can be seen from the discussion above a combination of adverse conditions, occurring simultaneously, could have a disastrous impact on a tailings basin water balance. This is demonstrated on Table 3 where the annual inflows are summarized for the base case and reasonable best and worst case scenarios. For the worst case scenario, the net annual inflow increases by a factor of over 3. This clearly demonstrates the wide range over which a tailings facility might have to operate in order to prevent unscheduled spills and for the system to stay in balance on an annual basis. For our example, it could be argued that the range could be from 0.78 Mm³/yr. to 4.53 Mm³/yr.

Table 3. Typical Variation In Net Annual Inflow

	Annual Inflows		
	Reasonable Best Case Scenario (Mm³)	Average Base Case (Mm³)	Reasonable Worst Case Scenario (Mm³)
Inflows			
Tailings Water	0.73	1.34	3.28
Mine Water	0	0	0.55
Run-off	1.35	1.80	2.26
Total	2.08	3.14	6.09
Losses			
Retained in tails	0.40	0.40	0.40
Seepage	0.03	0.03	0.03
Evaporation	0.14	0.14	0.14
Recirculation	0.73	1.20	0.99
Total	1.30	1.77	1.56
Net Annual Inflow	0.78	1.37	4.53
Parameters			
Tailings Production (t/day)	3,000	3,000	3,000
Slurry density (%)	60	45	25
Mine water (m³/day)	0	0	1500
Run-off (%)	-25%	avg.	+25%
Recirculation (%)	100	90	30

6 ZERO DISCHARGE CONDITIONS

Zero discharge to the environment is attainable in relatively dry climates, and may even be necessary if water has to be conserved.

Zero discharge can be achieved if:

1. the run-off component is small (either a dry climate or a small watershed)
2. the recirculation to the mill is high (approaching 100%), and
3. the volume of tailings is large enough to store the run-off component in the pore spaces of the tailings

7 WATER DISPLACED BY TAILINGS

Normally a tailings pond remains in balance on an annual basis; that is the total net annual inflow is decanted each year. However, a situation may exist where a pond has to be decreased in volume. For example, an existing body of water may have to be completely or partially displaced over the life of a mine or even in a shorter period to make room for tailings. This can be simply handled by discharging a portion of the existing water with the net annual inflow (calculated by WATBAL). In this case, as tailings are deposited water has to be displaced. As the water, retained in the pore spaces of the tailings is already accounted for in the balance, the volume displaced equals the total space occupied by the tailings slurry (solids plus water). This can be calculated merely by:

$$\frac{dry\ weight\ of\ tailings}{dry\ density\ of\ deposited\ tailings}$$

It is noted again that because the pore water is already accounted for separately in the water balance, the volume of water displaced is equal to the total deposited mass (tailings solids plus pore space) and not just the water displaced by the tailings solids.

8 STORM FLOWS

As is discussed above a tailings pond is sized for average base case conditions with an appropriate factor of safety plus an allowance for storm flows. Because tailings basin watersheds are small, storm flows are normally easy to handle.

One approach for storm flows, is to use two design floods; an environmental design flood (EDF) and a dam design storm (DDF). The EDF is typically a storm with a finite return period of say 20 to 100 years, possibly in conjunction with the maximum snow melt. This flood is retained and managed with the normal accumulation determined by WATBAL. Floods in excess of the EDF are either allowed to spill unimpeded (very large storms) or spilled slowly with a reduced retention time. In many cases the two storms may even be the same if spillage cannot be tolerated.

Under no circumstances can tailings be allowed to escape from a basin or a dam be allowed to overtop, therefore the DDF must be chosen to ensure that this does not happen. The routed DDF is used to establish freeboard and to size the emergency spillway. For a tailings basin it is frequently based on the largest possible storm resulting from the Probable Maximum Precipitation (PMP) which might or might not happen in conjunction with the maximum snow melt.

9 DISCUSSION AND CONCLUSIONS

WATBAL is merely a mathematical tool which adds and subtracts inflows and losses to a system. No special skills are required to operate the program or even to set one up. The results are only as good as the numbers that are put into it. WATBAL has proven to be reliable in predicting pond elevations and effluent flows from tailings basins. It is easy to use and more importantly it requires very little computer capacity. The input data are entered into the program in easily recognizable terms which facilitates changes and enables sensitivity analyses to be carried out easily and quickly.

An important conclusion, which can be reached from the hypothetical example analysed in the paper, is that a tailings effluent management system must be able to operate over a large range of operating and hydrological conditions. There is no such thing as an unique water balance for a tailings basin. As is demonstrated in the paper, the net annual inflow for a reasonable set of worst case parameters can be easily greater than three times the average base case flow.

10 ABBREVIATIONS

Abbreviations used in this paper are as follows:
    M - mega or million; m - metre; mm - millimetre; and, t - metric ton.

REFERENCES

Welch, D.E. and Firlotte, F.W. 1989. "Tailings Management in the Gold Mining Industry". Proceedings Vol. 14, International Symposium on Tailings Effluent Management. Halifax, Aug. 1989. The Metallurgical Society of the Canadian Institute of Mining and Metallurgy.

Welch, D.E., Botham, L.C., and Bronkhorst, D.   "Tailings Basin Water Management, A Simple Effective Water Balance". Environmental Management for Mining, 1992 Saskatchewan Conference, Saskatchewan Mining Association.

Tailings & Mine Waste'95 © 1995 Balkema, Rotterdam, ISBN 90 5410 526 7

Integrated design of tailings basin seepage control systems using analytic element ground-water models

Ray W.Wuolo
Barr Engineering Company, Minneapolis, Minn., USA

Paul E.Nemanic
Rhône-Poulenc, Inc., Cranbury, N.J., USA

ABSTRACT: Ground-water remediation and seepage interception typically involves the design of drains, trenches, and/or pump-out wells. The usual approach taken is to design a remediation system at one or more "trouble spots" without considering the hydrogeologic effects of the entire facility, the interaction of multiple remedial systems, or the influence of regional ground-water flow. A detailed ground-water flow model of an entire facility and surrounding area is an ideal tool for managing seepage interception and remediation. However, such a model is typically costly to develop and difficult to modify. The Analytic Element Method (AEM) of ground-water modeling is a relatively new approach to modeling ground-water flow and solute transport over very large areas. The AEM approach does not require the construction of a grid mesh and does not use conventional model boundaries. These attributes allow a ground-water flow model to be cost-effectively constructed and calibrated for an entire facility, covering thousands of acres with multiple tailings basins. An AEM model was developed for the Rhône-Poulenc trona facility in Green River, Wyoming. The model encompasses an area of approximately 25 square miles and includes several tailings basins, the Green River, fractured rock, meander-belt alluvium, and a number of existing remediation systems. The model was calibrated to observed hydraulic head conditions and then used to design a comprehensive seepage interception system that can be integrated in with the existing systems. The lack of a grid mesh allows for the detailed examination of flow near the tailings basins and remediation systems without sacrificing the big picture of ground-water flow over the entire area.

1 INTRODUCTION

Tailings pond seepage that enters the ground water ceases to become a tailings basin design challenge and becomes a ground-water clean-up and containment problem. The influence of leakage from the tailings basins is only one part of the problem. Local and regional geology, precipitation, production well pumping, and streams all contribute to the direction and rate of ground-water flow and migration of the contaminants from the basins. The complexity and scale of the influences on ground-water flow may be best addressed by evaluating ground-water flow over the entire facility, rather than in a small area adjacent to a leaking tailings basin. A facility-wide analysis may demonstrate that ground-water contamination emanating from many tailings basins can be intercepted by a relatively simple and inexpensive system, rather than many small, independently operated remediation systems. Such is the case at the Rhône-Poulenc trona (sodium bicarbonate) mine near Green River, Wyoming.

121

Trona tailings are slurried into several unlined tailings basins at the Rhône-Poulenc facility. A network of monitoring wells has disclosed that ground water containing high concentrations of total dissolved solids (TDS) and chemical characteristics similar to decant water in the tailings basins is migrating toward the Green River. Over the years, various drains and pumping wells have been installed in order to capture this seepage before it reaches the Green River. New areas of contamination are periodically disclosed, requiring the addition of more drains or wells. However, little consideration had been given to how these systems perform in concert or what area of seepage is intercepted by each system. A ground-water flow model of the entire facility and surrounding area was constructed in order to evaluate the effectiveness of the existing seepage interceptor systems and to design a system which integrates the various drains and wells.

2 FACILITY SETTING

2.1 *Facility Description*

The Rhône-Poulenc trona mine consists of underground mine works, a processing plant, and tailings ponds (Fig. 1). The plant workings are located on a bedrock bench adjacent to the meander-belt flood plain of the Green River. The tailings basins are located in upland areas that were previously incised by ephemeral streams, forming topographic draws. Tailings Pond 1 is situated in a former playa lake depression near the mouth of Stevens Draw. Tailings Pond 1 was formed by the construction of the Main Dam (earthen) across Stevens Draw and by the construction of the South Dam near the end of Chris Wash (topographic depression). Pond 2, which is no longer in use, is situated in a playa lake basin north of the plant. The smaller evaporation ponds are located in a shallow valley adjacent to Stevens Draw, northwest of Pond 4. Pond 4 is located behind Dam 2, approximately 1.5 miles east of the Main Dam and Stevens Draw.

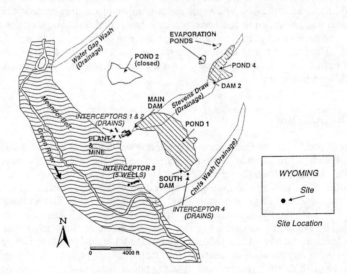

Figure 1. Facility and topographic features at the Rhône-Poulenc trona mine near Green River, Wyoming.

2.2 Topography and Physiography

The Green River is the dominant hydrologic feature. The Green River can best be described as a braided stream on the local scale and on a larger scale as a meandering stream. The typical surface water elevation of the Green River is approximately 6188 feet above mean sea level and the river flows south-southeast at a gradient of approximately 0.001 ft/ft.

The ancestral Green River built several terraces, three of which are present in the plant area. The lowest terrace is 10 to 15 feet above the river level and is locally referred to as the "Meander Belt" alluvium. The plant facilities are built on the middle terrace, locally referred to as "Stevens Flat". The third terrace is at elevation of approximately 6250 feet above mean sea level. The terraces can be thought of as steps in the bedrock, carved by the ancestral Green River, covered with a thin veneer of alluvium, as shown on Figure 2.

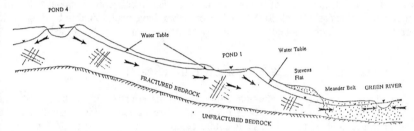

Figure 2. Schematic cross section through the tailings basins showing ground-water flow paths.

2.3 Geologic Setting

The bedrock at the plant site is the Eocene Bridger Formation, which overlies the Green River Formation. The Bridger formation consists of fluvially deposited sandy, tuffaceous mudstones and sandstones, with lacustrine shales and limestones.   The bedrock is exposed in cliffs and ledges and is a bench former in the vicinity of the plant.

The Bridger Formation is typically fractured to some extent to a depth of approximately 60 feet and fracturing may be quite pervasive to a depth of 50 feet. The depth of fracturing is related to the topography and is partially the result of stress relief due to the erosion of the upper strata. Along drainages and valley walls, vertical and diagonal fractures and weathering along fracture plains are visible.

2.4 Ground-Water Flow

The Rhône-Poulenc facility is in a semi-arid area that receives very little precipitation. Regional ground-water flow originates in the regional recharge areas along the south slope of the Wind River mountains, 20 to 30 miles north and northeast of the plant. Ground water flows in bedrock units and discharges to the Green River.

Ground water is present in two primary hydrostratigraphic units in the vicinity of the plant: (1) in the fractured upper portion of the Bridger Formation and (2) in the Meander Belt alluvium. An extensive network of monitoring wells has identified a continuous piezometric surface across the facility. Ground water flows from northeast to southwest in the fractured bedrock and discharges into the Meander Belt alluvium, where flow is roughly south and parallel to the surface gradient of the Green River.

2.5 *Ground-Water Quality*

Ground-water quality at the facility has been impacted by seepage primarily from Tailings Pond 1 and to a lesser extent from former Tailings Pond 2. The water in the tailings basins contains dissolved sodium and bicarbonate (the constituents of trona) at concentrations of hundreds of milligrams per liter. White precipitates form on the bank sediments of the basins. The tailings basin water typically is brown.

Ground water from monitoring wells downgradient of the tailings basins (particularly Tailings Pond 1) displays the same brown color as the tailings basin water. Concentrations of sodium, calcium, sulfate, and bicarbonate in samples from wells near the tailings basins are similar to those measured in the tailings basin water. High total dissolved solids (thousands of parts per million TDS) concentrations and specific conductance values as high as 20,000 μmhos/cm from have been found to be diagnostic of ground water which has been impacted by seepage from the tailings basins.

3 PREVIOUS APPROACHES TO SEEPAGE INTERCEPTION

There have been four seepage interception systems constructed at the facility, as shown on Figure 1. Interceptor Systems 1 and 2 are buried drains located near the plant, downgradient of the Main Dam. Interceptor System 3 consists of 5 pumping wells in the Meander Belt Alluvium downgradient of Pond 1. Interceptor System 4 is a 400-foot long drain with sump located immediately downgradient of the South Dam of Pond 1. The interceptor systems were installed in response to the detection of high-TDS ground water in monitoring wells located downgradient of the tailings basins. The approach to installing the interceptor systems has generally been to provide interception of seepage near the point of detection of high-TDS ground water.

Some consideration had been given to the rate and direction of ground-water flow when designing these systems. However, little attempt was made in evaluating the potential effectiveness of each system before installation. For example, Interceptor System 3 was installed in the Meander Belt alluvium to capture high-TDS ground water that was entering the alluvium from the fractured bedrock near the present location of Interceptor System 3. No attempt was made at evaluating the capture zone of this system or the effect of this system on the other existing systems.

Ground-water quality monitoring in recent years has made it obvious that seepage from the tailings basins (particularly from Pond 1) was resulting in facility-wide water-quality problems. A facility-wide approach was needed in order to design an integrated system of interceptors that would prevent tailings pond water from reaching the Green River.

4 MODELING APPROACH

4.1 *Model Selection*

A facility-wide ground-water flow model was deemed necessary in order to (1) evaluate the effectiveness of the existing interceptor systems and (2) design new interceptor systems which would collect all tailings pond seepage before it reached the

Green River. The model needed to be capable of evaluating ground-water flow across the entire facility yet provide simulations over a small area in order to assist in design of new systems.

Finite-difference and finite-element models are capable of simulating both regional and small scale ground-water flow conditions. However, these models employ a grid mesh which dictates the level of available detail that can be included. In order to obtain sufficient detail for designing interceptor systems at several location, a very small grid spacing would be necessary. A small grid spacing leads to a very large number of equations to solve, making the model difficult to use and costly. The approach selected for this problem was the Analytic Element Method (AEM) of ground-water flow model, developed by Strack (1989).

The AEM method is capable of simulating non-uniform, steady-state ground-water flow in three dimensions. The method does not use a grid mesh. Rather, closed-form analytic functions (called analytic elements by Strack) are used to simulate various hydrogeologic effects such as aquifer inhomogeneities, rivers, ponds, wells, and drains. The analytic elements are simultaneously solved by the method of superposition. The flow solution is written as a summation of harmonic functions and particular solutions to the Poisson equation (Strack, 1987). Analytic elements are not obtained by integration but, instead, by conformal mapping. Once the solution is calculated, hydrogeologic parameters such as hydraulic head, discharge potential, average velocity, and leakage rate can be obtained for any location in the aquifer.

The major operational difference between AEM and finite methods of modeling is that the AEM model domain is infinite and the user is not constrained by aquifer boundaries (in the traditional sense) or by a grid mesh. This characteristic of AEM allows the user to examine ground-water flow at a very small scale, or at a regional scale without changing parameters, boundary conditions, grid, or resolving the system of equations. The absence of a grid mesh can save time and effort during model development in cases where the model domain must be expanded or the model must be extensively modified. Another difference between AEM and finite methods is that mass is always conserved in AEM.

4.2 Model Development

A conceptual hydrogeologic model was developed to define the parameters and assumptions to be used in the AEM model. The conceptual model of ground-water flow is schematically illustrated in Figure 2.

The modeled area encompasses both the fractured bedrock (over which the tailings basins are located) and the Meander Belt alluvium. These two hydrostratigraphic units were considered to not be separate aquifers but, rather, were considered to represent hydrogeologic inhomogeneities in a single aquifer system in which the water table is present. An important assumption made in developing the model is that the fractured bedrock acts as an equivalent porous medium. That is, the fractures are sufficiently interconnected as to cause ground water to flow in an manner similar to flow in an unconsolidated material. The high degree of both vertical fractures and horizontal bedding plains suggests that this is a reasonable assumption.

The aquifer base elevation was evaluated from several boring logs distributed over the facility. These logs provide information on the depth of fracturing. The base of the modeled aquifer system was established at the depth in which fracturing substantially diminishes. The boring log data suggest that the fracture depth is relatively uniform (approximately 50 feet) and the elevation of the base of the fractures follows the surface topography.

The hydraulic conductivity of the aquifer at various locations was evaluated from packer test data. The hydraulic conductivity of the fractured bedrock ranged from 0.3 to 0.7 feet/day. In some areas of the facility, the water table extends up into the alluvium and colluvium overlying the bedrock. Hydraulic conductivity values in the alluvial materials are higher than in the fractured bedrock. A specialized version of the Single Layer Analytic Element Model (SLAEM) was used to account for vertical stratigraphic changes in aquifer parameters. The hydraulic conductivity of the Meander Belt alluvium was obtained from pumping test data and determined to be 40 feet/day. The spatial variations in aquifer parameters (base elevation, hydraulic conductivity, and porosity) were modeled using double-root analytic elements. These double-root elements are shown on Figure 3.

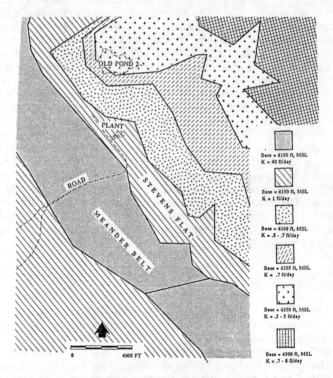

Figure 3. Double-root elements in AEM modeling which simulate changes in aquifer base elevation, hydraulic conductivity, and/or porosity.

The tailings ponds were considered to be the only important source of recharge to the aquifer over the facility, due to the arid climate. Ponds 1 and 4 were modeled using head and resistance-specified areal elements, which are not in direct hydraulic connection with the aquifer. The specified head used in the calibration of the model was the water-surface elevation in each pond. The resistance refers to the leakage resistance of the bottom sediments of the ponds (sediment thickness divided by vertical hydraulic conductivity). The resistance value was varied during the calibration process in order to obtain a good match between simulated hydraulic head and water levels measured in wells.

126

The Stevens Draw area between Pond 1 and Pond 4 contains several ground-water seeps. The hydraulic head in the aquifer is maintained at the ground-surface elevation of the draw. This feature was simulated in the model by using a line sink analytic element of specified head.

Interceptor Systems 1, 2, and 4 were simulated using drain analytic elements with hydraulic head specified at an elevation roughly equal to the invert of the drains. As part of the model calibration process, the simulated flow in the drains (calculated by the model) had to be nearly equal to the measured flow from the drain systems. Interceptor System 3 was modeled using five well analytic elements, with discharge in the model equal to that maintained in each pumping well.

5 SIMULATION OF EXISTING CONDITIONS

The AEM model was calibrated to (1) observed water-table elevations in monitoring wells and (2) measured discharges in the drain interceptor systems. Qualitative calibration checks were also performed to ensure that the ground-water flux into the Green River was approximately equal to base flow.

The calibrated steady-state ground-water flow conditions calculated by the AEM model are shown on Figure 4. The AEM model produces ground-water flow path traces at any point in the aquifer. The flow paths shown on Figure 4 were chosen to examine where seepage is being captured and where it is not.

Interceptor Systems 1 and 2 were found to be ineffective in capturing seepage from Pond 1. Most of the seepage water was found to flow underneath the drain systems of Interceptor Systems 1 and 2. Interceptor System 4 (drain) was found to capture very little of the seepage emanating from the South Dam area of Pond 1. Interceptor System 3, however, was found to be far more effective at capturing seepage that had been previously thought. The modeling results suggest that the higher permeability Meander Belt alluvium allows for substantially more water to be pumped than does the fractured bedrock. Once the seepage enters the Meander Belt alluvium, it flows south, parallel to the Green River, and is captured by the pumping wells of Interceptor System 3.

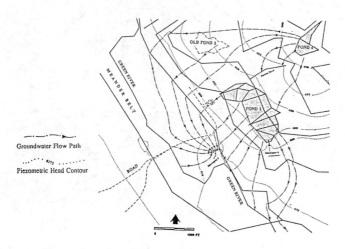

Figure 4. AEM model simulation of water-table elevations and ground-water flow paths for the existing seepage interceptor systems

6 DESIGN OF A COMPREHENSIVE INTERCEPTOR SYSTEM

The AEM model of the existing interceptor systems indicates that much of the seepage from the tailings ponds is captured by Interceptor System 3 before reaching the Green River. However, a substantial area south of Pond 1 is not being captured. Ground-water quality monitoring in the area south of the South Dam indicates that a high-TDS plume has moved approximately 1000 feet from the pond toward the Green River. This distance of migration is consistent with the time-of-travel computations performed by the AEM model. It is in this area that additional interceptor design is focused.

Several hypothetical interceptor systems were evaluated using the AEM model. The new system must capture seepage from Pond 1 without interfering significantly with the effectiveness of Interceptor System 3. Two main interceptor designs were examined: a system with wells and a drain/trench system.

The simulation of an interceptor system using wells is shown on Figure 5. The modeling results suggest that 12 wells, each pumping a 1 gallon per minute, spaced over a distance of approximately 3,000 feet would be sufficient to capture seepage from Pond 1. A drain/trench system simulation was also modeled. The modeling results suggest that a drain with an invert elevation of 6170 feet above mean sea level would capture seepage from Pond 1.

The drain/trench system, which is most attractive from a system operations stand point, was initially selected as the interceptor system alternative. However, this alternative was later rejected because of the difficulty in excavating to the design elevation. Design with the well system then proceeded.

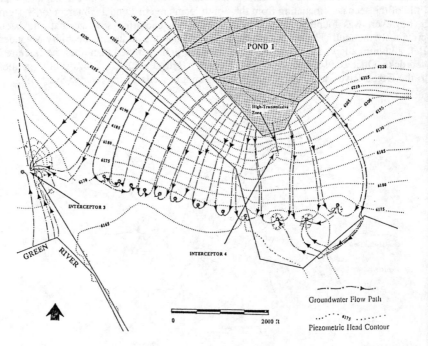

Figure 5. Predicted seepage capture for new well interceptor system consisting of 12 wells, each pumping at 1 gallon per minute

The model was used to design a pilot-scale system of five wells, located on the east end of the proposed well alignment by establishing the well spacing and predicting both the steady-state drawdown and the time required to reach steady-state conditions (approximately one year). During installation of the pilot-scale wells, a transmissive flow zone was discovered. The disclosure of the transmissive zone required further data collection, geophysical surveys, and re-adjustment of the model before proceeding with full-scale design. In addition, the effects of raising the dams in Pond 1 on the proposed interceptor system is being evaluated with the AEM model.

7 CONCLUSIONS

Interception of seepage from tailings basins should not be viewed in terms of a localized problem but, rather, in the context of regional ground-water flow. The effectiveness of individual systems, in relation to each other, is an important consideration when designing seepage interception systems. Ground-water flow modeling has long been considered an effective tool for such an endeavor but commonly used finite-method models have the drawback of a grid mesh which greatly affects the level of detail. Analytic Element Method (AEM) models do not use a grid mesh or conventional model boundaries. These models allow for detailed evaluation of individual interceptor system design as well as the effectiveness of the individual systems in the context of the entire facility.

REFERENCES

Strack, O.D.L. 1987. The analytic element method for regional groundwater modeling. *Proc. NWWA Conf. Solv. Ground Water Prob. Models.* 929-941. Dublin, Ohio.

Strack, O.D.L. 1989. *Groundwater mechanics.* New Jersey: Prentice-Hall.

Author index

Baker, F.G. 67
Botham, L.C. 109

Canali, G.E. 91
Canaly, C. 35
Chiu, T.-F. 1
Cotter, E.T. 11
Curry, J. 19

Eaton, W.S. 55
Ehrenzeller, J.L.

Filipek, L.H. 19

Gilmer, T. 35
Guo, X. 67

Harrison, W.J. 23, 43
Hathaway, C.J. 81
Hauff, P.L. 79

Johnson, J.M. 109

Kaszuba, J.P. 23
Kent, D. 35
Klco, K. 35

Livo, K.E. 79

McGaffey, K.M. 55
Mellott, W. 35
Moran, R.E., 55
Mueller, M. 35

Nemanic, P.E. 121

Papp, C.S.E. 19
Paschke, S.S. 43
Patton, C.A. 55
Pavlik, H.F. 67
Peters, D.C., 79

Runke, H.M. 81

Simons, D.B. 91
Simons, R.K. 91
Stern, J. 35
Stilwell, C.T. 105

Te-Fu Chiu 1
Ter Kuile, M. 35

VanWyngarden, T.J. 19
Voorhees, J.S. 67

Welch, D.E. 109
Wendtlandt, R.F. 23
Wildeman, T.R. 19
Wuolo, R.W. 121

Xiaoniu Guo 67